AF598391

TRANSITION METAL IMPURITIES IN SEMICONDUCTORS

TRANSITION METAL IMPURITIES IN SEMICONDUCTORS

E M OMEL'YANOVSKII

V I FISTUL'

M V Lomonosov Institute for Fine Chemical Technology, Academy of Sciences of the USSR, Moscow

Translated from the Russian by

ALBIN TYBULEWICZ

Editor, 'Soviet Physics — Semiconductors'

ADAM HILGER LTD, BRISTOL AND BOSTON

British Library Cataloguing in Publication Data

Omel'yanovskii, E. M.
Transition metal impurities in semiconductors.
1. Semiconductors—Defects
I. Title II. Fistul', V. I. III. Primesi perekhodnykh materialov v poluprovodnikakh. *English*
537.6′22 QC611.6.D4

ISBN 0-85274-493-5

First published in 1983 as *Primesi perekhodnykh materialov v poluprovodnikakh* by Metallurguia, Moscow

Published by Adam Hilger Ltd
Techno House, Redcliffe Way, Bristol BS1 6NX, England
PO Box 230, Accord, MA 02018, USA

Printed in Great Britain by J W Arrowsmith Ltd, Bristol

CONTENTS

PREFACE TO THE ENGLISH EDITION

Doping is the most widely used method for imparting the required properties to semiconductor materials. Therefore, a study of impurity centres and particularly of those with strong electron localization will for many years be one of the central problems in theoretical and experimental investigations of semiconductors.

We are grateful to Adam Hilger for undertaking to publish our book in the English language. In this way the Western reader will be able to become more fully acquainted with the work of the Soviet scientists on this topic. We hope that this will in turn help in further growth of the physics of semiconductors.

In the time since we wrote this book there have been quite a few investigations which we were unable to include. At the end of the bibliography we added a brief list of the most important, in our view, recent work from which the reader will gain an idea on the development of the topic during the period 1983-1985.

The appearance of the English edition has been greatly assisted by the helpful attitude of a number of English scientists: Prof. L.J. Challis and Dr. L. Eaves from the University of Nottingham, and Prof. A.K. Jonscher from Chelsea College of the London University, whose opinion we greatly value because they themselves made major contributions to the study of the behaviour of transition metal impurities in semiconductors.

It gives us great pleasure to acknowledge the high quality of the translation carried out by a major expert on Soviet literature on semiconductors Mr. A. Tybulewicz.

PREFACE TO THE RUSSIAN EDITION

The improvement in the quality of products and the development of new materials satisfying modern requirements plays an important role in the scientific and technical revolution. Consequently, the "Basic Directives on Economic and Social Development in the Soviet Union" for 1981-1985 and for the period up to 1990, adopted at the Twenty-Sixth Congress of the Communist Party of the Soviet Union, stress strongly the growth of theoretical and experimental research relating to the development and extensive adoption of new semiconductor materials and of electronic devices made from them.

A successful performance of this task largely depends on widening the range of impurities used in doping because this process still remains the main method for imparting the necessary combination of properties to semiconductor materials.

In the last 10-15 years the widely used (in the semiconductor technology) and relatively thoroughly investigated impurities forming "shallow" hydrogen-like levels in the band gap of the semiconductor matrix have been supplemented by a whole class of impurities creating "deep" local states with an ionization energy comparable with the band gap of a crystal.

For a long time the presence of deep-level impurities in semiconductor materials has been regarded as an undesirable harmful effect in semiconductor technology, which causes deterioration of the properties of the materials and of the quality of the devices. Such impurities do indeed frequently limit the attainable purity of many semiconductor crystals and create recombination capture centres, which reduce the quantum efficiency of the luminescence emitted by semiconductor sources of spontaneous and coherent radiation. The centres which have deep states usually migrate readily and are frequently responsible for the observed degradation of the various characteristics of semiconductor devices.

However, controlled introduction of deep-level impurities into crystals is being used actively to solve a number of problems important in the physics and technology of semiconductors. Substrates made of semi-insulating gallium arsenide and other III-V compounds, prepared as a result of doping these materials with deep-level impurities, are used extensively in modern microelectronics. Such crystals are also used successfully in optoelectronic devices, for example as infrared-radiation modulators, etc. The short lifetime in crystals doped with impurities makes it possible to construct devices with a very short response time. Doped photodetectors made of these materials and covering an extremely wide range of wavelengths are of major interest.

Even this brief, far from complete, enumeration of examples of practical applications of deep-level impurities demonstrates convincingly the importance of the problem of deep centres in the physics and technology of semiconductors.

Among these impurities a special place is occupied by transition (T) metals, the main distinguishing feature of which is the incompleteness of their d-shells which also determines the characteristic properties of these impurities in semiconductor materials. The results of experimental studies of T-metal impurities in semiconductors are scattered between many periodicals. A systematic account and analysis from a unified standpoint would obviously be useful to engineers, physicists, and chemists working on semiconductor materials science and in electronic technology.

The present book was written as follows: E.M. Omel'yanovskiĭ wrote the Introduction and Chap. 2, whereas V.I. Fistul' wrote Secs. 1, 2, 3, 7 in Chap. 1; Secs. 1 and 2 in Chap. 3, and the whole of Chap. 4. The remainder of the book was written by both authors.

The authors are grateful to researchers who supplied their data and discussed many aspects of the problems considered in the book. They will be very grateful to the readers for critical comments and suggestions.

INTRODUCTION

The majority of the remarkable properties of semiconductors that have determined the rapid growth of solid-state electronics are due to the presence in semiconductor crystals of impurity atoms in concentrations which can be varied in a controlled manner in an extremely wide range directly during growth or after preparation of the material by diffusion, ion implantation, etc.

Since the late forties an enormous effort has been made to prepare impurity-free germanium and silicon crystals. Currently the concentration of electrically active residual impurities in Ge and Si does not exceed 10^{10} and 10^{12} cm^{-3}, respectively, but materials of this purity have very limited direct applications. This applies also to GaAs crystals. The main achievement of the semiconductor technology is that the current level of purity of elemental semiconductors makes it possible to carry out deliberate precision doping of a material with specific impurities, thus imparting the required properties to the material, which – in their turn – determine the quality of the fabricated devices.

The problem of impurity centres encountered right at the beginning of the semiconductor era is still topical. On the one hand, the great majority of the existing devices are based on the use of crystals specially activated with impurities and having specified properties. On the other hand, our knowledge of the properties of semiconductor materials after introduction of dopants is essentially empirical. Consequently, the attempts to forecast the state and behaviour of many impurity centres frequently give inconsistent results and the values of the principal parameters (energies of the ground and excited states; photoionization, capture, scattering cross-sections of carriers; state of an impurity in a crystal lattice; g-factor; solubility and distribution coefficient, etc.) can be determined only experimentally.

Controlled introduction of impurity atoms makes it possible to create such physical situations in crystals which can cause radical changes even in stable fundamental properties of the semiconductor matrix. It is therefore not surprising that the

local centres in crystals have to a large extent become an independent topic for investigation in modern solid-state physics.

The problem of the impurity states has been at the centre of attention of theoreticians and experimentalists for more than two decades. Special attention has been given to centres of strong localization of carriers with the ionization energy of the order of the band gap. These centres are known as deep. In contrast, the local centres with an ionization energy E_i much less than the band gap are known as shallow.

This division into deep and shallow levels is in need of major refinement. In our opinion, it is more correct to regard as shallow those levels which can be described satisfactorily within the framework of the simple or generalized variants (Chap. 2) of the effective mass method. It will be clear from further discussions that, from this point of view, we shall regard as shallow impurities those deep "isocore" impurities (for example, sulphur in silicon) with the potential which is free (as in the case of hydrogen-like centres) of the short-range (non-Coulomb) part with all the simplifications which follow from this circumstance. We shall use the term "deep centres" in those cases which for fundamental reasons cannot be described in the effective mass approximation. The number of such impurities in semiconductors is very large.

The current theory of deep centres cannot provide a unified description of the majority of their properties, although it is deep centres that exert an important (and in some cases decisive) influence on the properties of semiconductors and of the majority of semiconductor devices. The main difficulties encountered in the theory are due to the fact that a deep centre represents a many-electron impurity atom and it experiences the action of a field created by atoms in the crystal lattice, the influence of which can no longer (in contrast to shallow hydrogen-like centres) be allowed for simply by introducing an effective mass and a macroscopic permittivity of the matrix in the Schrödinger equation. Therefore, in spite of the numerous attempts to construct a satisfactory theory of deep centres in semiconductors, the experimental observations are usually explained by semi-empirical models.

An analysis of the current status of the theory of local centres can be found in the well-known reviews of Bassani, Iadonisi and Preziosi, Roĭtsin, and Pantelides. Therefore, in the present monograph we shall consider briefly only those investigations which are deemed to be most interesting for the development of the ideas on the centres of strong localization that are necessary for the understanding of their specific properties and of their influence on the principal parameters of semiconductor materials observed experimentally.

Four main approaches to the theory of deep levels in semiconductors are employed at present: the generalized effective mass method, the pseudopotential method, the Green function method, and the quantum-chemical methods. They will be considered in this sequence in our book.

It has been known for some time that impurities in the form of atoms of transition metals with a partly filled d-shell (T-metals) are centres of strong localization in germanium, silicon, and other semiconductor materials, and they have a particularly strong influence on the nonequilibrium properties of these materials. Studies of the properties of these semiconductors with transition metal impurities have been followed mainly along two avenues. The first avenue is the approach limited to measurements of the energy positions of the relevant deep levels and their cross-sections for the capture of nonequilibrium carriers, governing the recombination characteristics of the doped materials. The second avenue is based on consideration of the nature, symmetry of the wave function of the introduced centre, and parameters of the spin Hamiltonian describing the interaction of an unpaired electron with the ion core of a host atom and with the crystal environment. It should be stressed that these two ways of approaching the problem have been pursued independently and there have been practically no attempts to compare the results.

Systematic investigations of the state and behaviour* of impurities with a partly filled d-shell in III-V compounds have been carried out mainly because of the need to develop semi-insulating GaAs crystals which are used widely in modern electronics. However, in spite of this purely practical approach, the interest in the study of the properties of T-metal impurities in III-V compounds arises also from the need to develop further the basic ideas on deep centres in solids.

The incomplete nature of the d-shell in transition metals

---

*According to the established terminology, the state of an impurity is usually understood to be the structure of a centre, its nature, charge states, and local symmetry in the host crystal lattice; the term behaviour of an impurity covers the energy spectrum, carrier- and photon-capture cross-sections, degeneracy factor, and other parameters of the centre governing the characteristic properties of the investigated system as a whole.

is clearly the main reason for the strong localization of electrons at these impurity centres in the majority of semiconductors. We shall show in the present monograph that there are many other reasons for considering transition metal impurities as a suitable model system of deep centres in crystals, which is so essential for the development of a consistent theory. In this connection it is appropriate to mention that in order to describe the properties of centres with deep levels of different origin (such as oxygen in III-V compounds, or group VI elements in elemental semiconductors, or vacancies in compounds) it may be necessary to modify significantly the ideas put forward below.

A favourable circumstance for a detailed study of the properties of transition metal impurities is the fact that ions with a partly filled d- or f-shell retain a finite net momentum in the crystal lattice. The paramagnetism of these centres makes it possible to utilize effectively the powerful arsenal of resonance methods for investigating the state of impurities introduced into a semiconductor matrix. It is these methods that have made it possible to determine, in particular, the charge configurations of the ions belonging to the iron group and to show sufficiently convincingly that in wide-gap III-V compounds they form a substitutional solid solution at the cation sublattice sites, in contrast to elemental semiconductors where these impurities may be located also at interstices.

Out of the great variety of the properties of T-metal impurities in semiconductor crystals, we shall concentrate our attention on the following topics: the state of transition metal impurities with a partly filled 3d-shell in the lattices of germanium, silicon, and some III-V compounds; the energy spectrum and the main parameters of deep local states formed by these impurities in crystals; the influence of all these factors on the electrical, optical, photoelectric, and magnetic properties of the investigated materials.

Some of the important properties of crystals doped with transition metal impurities are not discussed in our book. For example, for reasons of space, we have been unable to analyse the diffusion phenomena in semiconductors containing transition metal impurities and to deal with the greatly extended applications of these phenomena in semiconductor devices. However, the characteristic features of the state and behaviour of impurity centres with a partly filled 3d-shell considered in the present monograph provide an opportunity not only to interpret qualitatively the known properties of materials, but also to forecast the main characteristics of the hitherto not investigated systems to deliberately modify their properties over a wide range. Our view on future trends in the development of our knowledge of the properties of transition metal impurities in other deep centres in semiconductors will be presented in a brief resumé at the end of the book.

1. STATE OF TRANSITION METAL IMPURITIES IN CRYSTAL LATTICES OF SEMICONDUCTORS

1.1. CRYSTAL CHEMISTRY OF SEMICONDUCTORS WITH TRANSITION METAL IMPURITIES

The traditional approach used to determine the states of impurity atoms in semiconductors has been via crystal chemistry. The starting point is the crystallochemical correspondence (or the lack of it) between the impurity atoms and the positions which they can (or cannot) occupy in a crystal. The crystallochemical correspondence principle implies that one should consider simultaneously the size and electron aspects of this correspondence.

The size correspondence means that the stability conditions of a crystal structure should be satisfied. This is possible only for specific ratios of the size of the impurity atom and the position it occupies. The positions in a crystal can either be sites occupied by the host atoms and in this case we speak of substitutional solid solutions or the voids (interstices) in the crystal lattice and in this case the solid solutions are known as interstitial.

The electron correspondence implies that a stable chemical binding can be established between an impurity atom and the surrounding atoms, which depends on the nature of the position occupied by the impurity atom.

Unfortunately, the state of the theory is such that in most cases it is necessary to consider the two types of correspondence separately, which is the main reason why it is often not possible to make forecasts on the basis of the crystallochemical approach. It is sufficient to point out that the dimensions of impurity atoms themselves depend very strongly on the nature of the chemical binding. Therefore, there are various kinds of radii of atoms: covalent radii, ionic, tetrahedral, radii of atoms in compounds, etc.

Nevertheless, the concept of the crystallochemical correspondence is often valuable. We shall adopt it mainly to demonstrate those features which appear when the impurities are atoms of transition metals with a partly filled d-shell.

We shall begin by considering the size correspondence and we shall do this by turning to the well-known theory of close-packed structures because this approach makes it possible to determine all the necessary geometric characteristics of crystal structures of group IV semiconductors and of III-V semiconductor compounds. In this theory the formation of crystals is regarded as successive formation of layers of hard spheres. By way of example (Fig. 1), we shall consider the case when any three spheres A in the first layer, packed closely on a plane, form recesses which can accommodate closely packed identical spheres B of the second layer. When the third layer is formed there are two possibilities: these are illustrated in Figs. 1a and 1b. In the former case the spheres in the third layer occupy recesses in the second layer or – which is equivalent – they are located above the spheres in the first layer, whereas in the latter case they are above the spheres of the second layer, i.e. above the recesses in the first layer.

In the fourth layer there are again two possibilities of placing the spheres above those in the third layer (Fig. 1c) or the second layer (Fig. 1d). Clearly there are many ways of distributing the spheres in the layers. For our purposes the sequence of the ABCABC type is important because it corresponds to packing in a unit cell of a face-centred cube (Fig. 2a), which is the crystal structure of germanium and silicon, and of III-V semiconductors. When a sphere is placed in a recess formed by three spheres on a plane in the preceding layer, the centres of all four spheres form a tetrahedron shown in Fig. 2b.

Simple geometric considerations can thus be used to obtain all the quantities representing the tetrahedron and the unit cell in the face-centred cubic (f.c.c.) lattice. These quantities are in one-to-one correspondence with the radius of the spheres in a layer, i.e. with the radius of the host lattice atoms and, consequently, with the period of the crystal lattice a; they are listed in Table I.

In addition to the tetrahedral voids or interstices, the f.c.c.

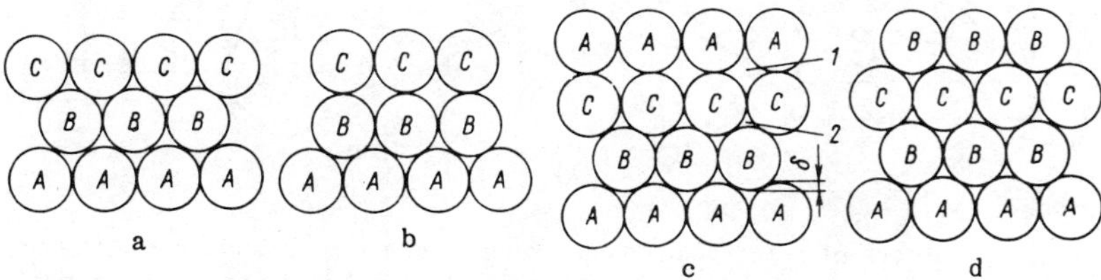

FIG. 1. Some types of packing, illustrating formation of tetrahedral and octahedral interstices (voids): 1) octahedral interstice; 2) tetrahedral interstice; δ is the depth of penetration of spheres into recesses.

TABLE I. Relationship Between Quantities Representing Unit Cell and Tetrahedron in Face-Centred Cubic Lattice with Atomic Radius in Lattice Period

Parameter	Designation	$f(R)$
Crystal lattice period	a	$a = 2R\sqrt{2}$
Edge of tetrahedron	a_1	$a_1 = 2R = 0.71a$
Height of tetrahedron	h	$h = 2R\sqrt{2/3} = 0.578a$
Length of unit cube diagonal	D	$D = 3h = 4R\sqrt{3/2} = a\sqrt{3}$
Distance from centre of tetrahedron to its base	Z	$Z = R/\sqrt{6} = 0.41R = 0.145a$
Distance of centre of tetrahedron to vertex	y_T	$y_T = h - Z = (3/\sqrt{6})R = 0.433a$
Depth of penetration of spheres into recesses	δ	$\delta = 2R - h = 0.367R = 0.13a$
Distance of centre of octahedron from vertex	y_O	$y_O = R\sqrt{2}$
Minimum radius of sphere inscribed in tetrahedral interstice	r^T_{min}	$r^T_{min} = y_T - R = 0.225R$
Minimum radius of sphere inscribed in octahedral interstice	r^O_{min}	$r^O_{min} = y_O - R = 0.414R$
Maximum radius of sphere inscribed in tetrahedral interstice	r^T_{max}	$r^T_{max} = 0.5R$
Maximum radius of sphere inscribed in octahedral interstice	r^O_{max}	$r^O_{max} = 0.732R$

lattice contains also octahedral voids. They appear when a recess between three spheres in one layer coincides with a recess between three spheres in the next layer, displaced by 60° relative to the first layer. An octahedral interstice has the coordination number 6. Its centre is separated from each of the nearest six spheres by a distance y_O, which can also be related to the sphere radius R on the basis of geometric considerations (Table I). There are four spheres per unit cell of the f.c.c. lattice and, consequently, there are four octahedral interstices and eight tetrahedral interstices.

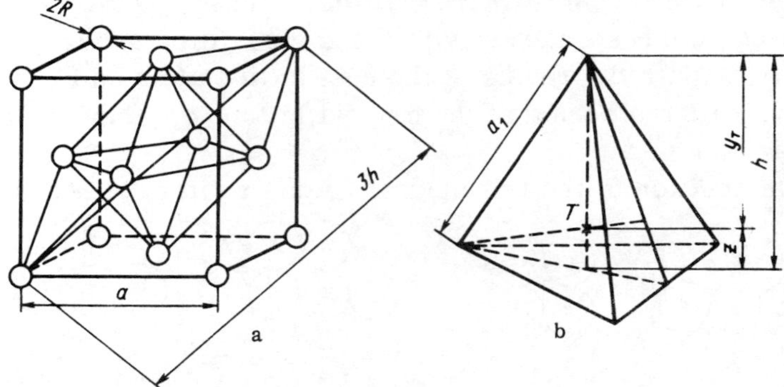

FIG. 2. Unit cell of a face-centred cube (a) with the period a and a unit tetrahedron formed in the close packing of spheres (b). Here, T is the centre of the tetrahedron, a_1 is the edge of the tetrahedron, $h = y_T + z$ is the height of the tetrahedron.

The real crystal lattice of the investigated semiconductors is shown in Fig. 3. The tetrahedral and octahedral interstices are distinguished in this figure. The centres of the tetrahedral interstices T are located on the body diagonals of the cubes, half-way between two unbound atoms at a distance y_T from each of the four surrounding atoms. The centre of an octahedral interstice O is located at the midpoint of a cube edge and is separated from each of the nearest six atoms by y_O.

Germanium and silicon crystals contain atoms of just one kind. Consequently, there is only one type of tetrahedral interstices and one type of octahedral interstices in the lattices of these materials. A lattice of a III-V crystal consists of two sublattices

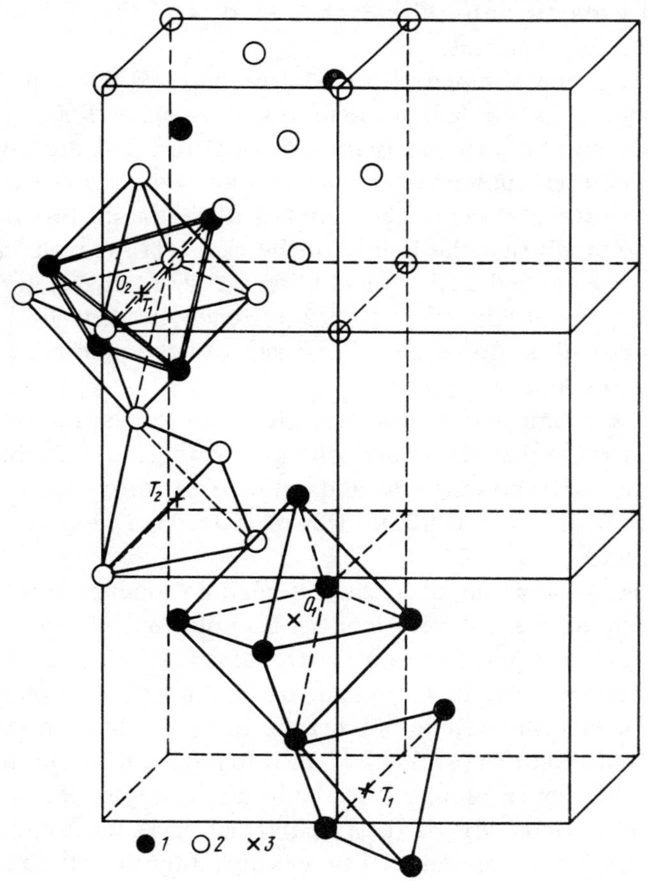

FIG. 3. Diamond-like crystal lattice of a III-V semiconductor: 1) A cations; 2) B anions; 3) centres of tetrahedral and octahedral interstices.

formed from the group III and V atoms (they are identified by the black and open circles in Fig. 3, respectively). Such a lattice has two types of tetrahedral interstices T_1 and T_2, and two types of octahedral interstices O_1 and O_2, which differ in respect of the nature of the surrounding atoms.

The geometric characteristics of tetrahedral and octahedral interstices make it possible to define the radii of the spheres r^T_{min} and r^O_{min}, into which they can be inscribed. Their values, also listed in Table I, are the minimum possible dimensions of the inscribed spheres, because when the radius of such a sphere is smaller the system becomes unstable. This was postulated by Magnus and is widely used in crystallography. However, the number of exceptions to this Magnus rule is so large that its validity is quite doubtful.

There are also upper limits to r^T and r^O, i.e. the radii of the inscribed spheres cannot assume values exceeding r^T_{max} and r^O_{max}. They can be deduced from the condition that the layers in a crystal should not lose contact with one another. Otherwise the effective dimensions of the spheres of the basic layer would increase so much that the height of the new tetrahedron h' would be equal to or exceed 2R. Since $h' = 2R'\sqrt{2/3}$, it follows that for $h' = 2R$, we obtain $R' = R\sqrt{3/2}$. Hence, we can find the maximum permissible values of the radii of the spheres inscribed into the interstices r^T_{max} and r^O_{max}.

Such a geometrical analysis allows us to assume (if we ignore the Magnus rule) that the conditions governing the distribution of impurity atoms in tetrahedral and octahedral interstices are of the form: $r \leq 0.5R$; $r \leq 0.732R$, where r is the radius of an impurity atom.

We recall that impurity atoms need not occupy interstices and can form also a substitutional solid solution. In this case there is a greater justification for the Magnus rule. In fact, we shall assume the core of a crystal to consist of ions (ionic crystals) or of atoms with the valence electrons in the collective state (covalent and metal crystals). If an impurity ion is smaller than the ions in the environment, the Coulomb energy of the interaction will be weak. However, if the impurity atom is not ionized, its binding to the environment will be governed by the elastic forces and in this case the binding energy (which decreases with the distance as 1/r) is also low. Consequently, the position of a small impurity atom at a regular site in the lattice becomes unstable.

All this is more or less true also of the localization of the wave functions of the electrons of an impurity atom within the first coordination sphere. If the wave function extends over a larger number of the coordination spheres, the coupling with the more distant atoms in the environment increases the stability of a small atom at a substitutional position. In this case we can expect formation of substitutional solid solutions also by atoms with $r < R$. This case is realized, for example, in boron-doped silicon crystals.

In the opposite case when the radius of an impurity atom is slightly greater than the radius of the atom it replaces, the substitution is still possible. The size condition for the replacement then becomes $r \approx R$. This size approach gives results in reasonable agreement with the experimental data in the case of ionic crystals and specifically in respect of the distribution of impurity ions in the lattices of such crystals. Here R and r are the ionic radii, tables of which have been compiled on the basis of the empirical data by Goldschmidt, from "theoretical" data by Pauling, from mixed experimental and theoretical data by Ahrens and Kordes, and from crystallochemical data by Bokii and others. In the case of atoms of some of the elements the values of the ionic radii given in such tables are very different. Therefore, in the majority of cases it is extremely difficult to select the ionic radii necessary in the analysis. It is worth noting the attempt by Batsanov[1] to construct a system of radii which vary continuously as a function of the ionicity of a given compound (or crystal) in which the investigated atom (or ion) is located. Figure 4 shows curves illustrating changes in the size of the atoms of germanium, silicon, and some group III and V elements on increase in the degree of their ionization (ionicity), plotted by us on the basis of the results of Batsanov.[1] All the values of r in this figure apply to the tetrahedral coordination when the number of ligand atoms around the atom under consideration is four.

Conversion of radii to the case of a different number of ligands can be made using the factor $\beta = (L/6)^{1/n-1}$, where n is the average Born power exponent for anions and cations, which depends to which rare gas or to which noble metal ion does the electron configuration of the investigated ion correspond.[2] In the case of aluminium, silicon, and phosphorus ions this is the neon structure and we have $n = 7$; in the case of scandium, titanium, vanadium, chromium, manganese, iron, cobalt, nickel, gallium,

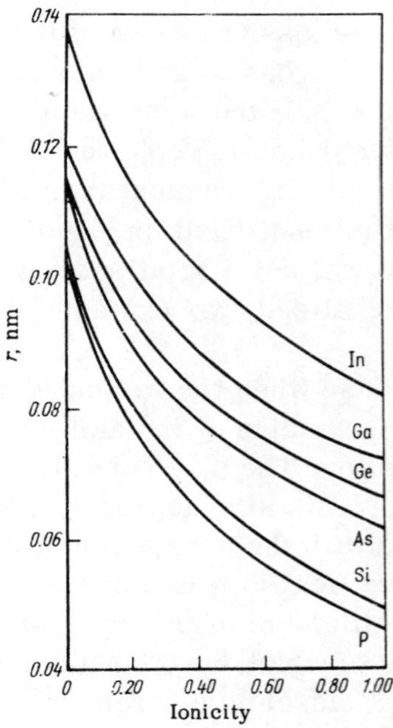

FIG. 4. Dimensions of atoms plotted as a function of the degree of ionization.

germanium, and arsenic – this is the structure of argon and we have n = 9, whereas for indium and antimony the correspondence is to the krypton structure and n = 10.

The correction factors β for cases when the number of ligands differs from four are listed below:

L	4	6	8	12
β for n equal to:				
6	1.000	1.084	1.148	1.245
9	1.000	1.051	1.090	1.147
10	1.000	1.046	1.079	1.130

Obviously, in crystallochemical speculative analysis one has a wide range of values of r to choose. Only in the case of the semiconductors germanium and silicon it is possible to select more or less reliably the value of r_0 corresponding to zero ionicity and this can be done because of the basically covalent nature of the semiconductors. In the case of III–V crystals the covalent type of chemical binding is accompanied by a contribution of ionicity, but it is difficult to determine the degree of ionicity although attempts have been made repeatedly.

Table II lists the limiting values of r converted by us to the

TABLE II. Ionic Radii of Silicon, Germanium, III and V Elements, and 3d Transition Metals

Element	r_0	r_i	r^T_{mo}	r^T_{mi}	r^O_{mo}	r^O_{mi}	Element	r_i	
								r_i =4	n=6
Si	1.06/0.84	0.49/0.39	0.53/0.42	0.24/0.19	0.78/0.62	0.36/0.28	Sc	0.75/0.89	0.79/0.81
Ge	1.16/0.89	0.65/0.50	0.58/0.44	0.32/0.25	0.85/0.65	0.47/0.37	Ti	0.65/0.64	0.69/0.68
Al	1.18/0.95	0.58/0.47	0.59/0.47	0.29/0.23	0.86/0.69	0.42/0.34	V	0.59/0.56	0.62/0.59
Ga	1.19/0.99	0.71/0.59	0.59/0.49	0.35/0.29	0.87/0.72	0.52/0.43	Cr	0.55/0.49	0.58/0.52
In	1.37/1.30	0.81/0.77	0.68/0.65	0.40/0.38	1.00/0.95	0.59/0.56	Mn	0.52/0.44	0.55/0.46
P	1.03/0.74	0.45/0.32	0.50/0.37	0.21/0.16	0.75/0.54	0.33/0.23	Fe	0.71/0.61	0.73/0.64
As	1.16/0.84	0.61/0.44	0.58/0.42	0.30/0.22	0.85/0.61	0.45/0.32	Co	0.68/0.60	0.72/0.63
Sb	1.32/1.23	0.69/0.64	0.66/0.61	0.34/0.32	0.97/0.90	0.50/0.47	Ni	0.68/0.60	0.72/0.63

Notes. 1. The numerator gives the maximum values corresponding to the Batsanov ionic radii and the denominator gives the minimum Pauling values.

2. Designations used in above table: r_0 is the atomic radius; r_i is the ionic radius for 100% ionicity; n is the number of ligands in the environment; r^T_{m0} and r^T_{mi} are the maximum radius of a sphere which can be inscribed into a tetrahedral interstice formed by atoms or ions of a given element; r^O_{m0} and r^O_{mi} is the maximum radius of a sphere inscribed in an octahedral interstice formed by atoms or ions of a given element.

case of the four-ligand environment. All the other values published in the literature lie between those listed in Table II. This table gives also the maximum permissible sizes of the impurity atoms or ions which can be placed in tetrahedral and octahedral interstices in silicon, germanium, and III-V compounds. Table II provides also the values of the ionic radii of the 3d transition metals in the four- and six-ligand environments. We must bear in mind that these radii represent ions in the charge state Me^{3+}. A comparison of the radii makes it possible to determine the feasibility of placing transition metal impurities in different crystallographic positions in semiconductors. We shall consider the results of such a feasibility study in comparison with the experimental data for all group IV semiconductors and III-V compounds. The vertical shading in Fig. 5a represents the range of the values of the ionic radii of transition metals from the maximum to the minimum.

The covalence of the chemical binding in a silicon crystal makes it possible to compare the values of r_j of transition (T) metals with the atomic and not the ionic radius of silicon, which is represented by oblique shading in Fig. 5. The same figure

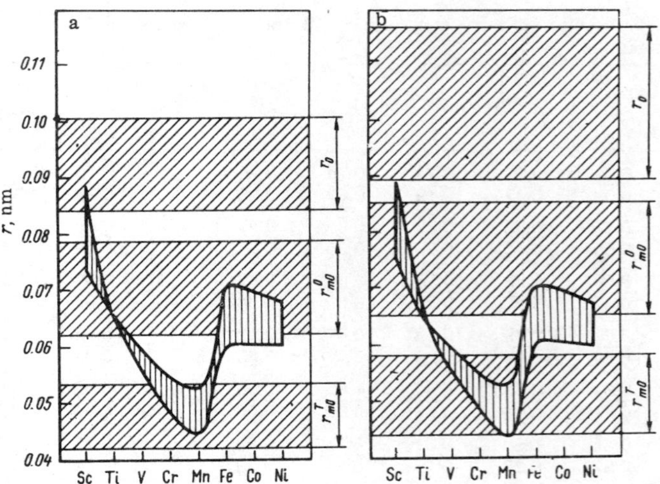

FIG. 5. Radii of transition metal ions and possible positions in the crystal lattices of silicon (a) and germanium (b).

shows the ranges of the values of r^T_{m0} and r^O_{m0} which represent the dimensions of tetrahedral and octahedral interstices formed by the atoms of silicon. We can see that apart from scandium, none of the other T-metals satisfies the condition $r \approx R$, i.e. in this approximation T-metal atoms cannot form substitutional solid solutions in silicon.

It follows from the data in Fig. 5 that some of them (vanadium, chromium, manganese) have radii more suitable for the location in tetrahedral interstices, whereas scandium, titanium, iron, cobalt, and nickel are more suitable for octahedral interstices. Practically the same comments apply also to germanium (Fig. 5b). In this case the formation of the substitutional solution of scandium is unlikely.

A different situation is found in the case of the T-metal impurities in III-V crystals. These crystals cannot be regarded as completely covalent. For the estimates we shall regard them as having 100% ionicity, which is of course a rough approximation. Some justification is provided by the good results obtained in calculations of the energy band structure of III-V compounds on the basis of this approximation. Moreover, any assumption about the degree of ionicity of III-V compounds is no better and not more justified than that just adopted. When the 100% ionicity hypothesis is made, a comparison of the relevant radii gives the results shown in Figs. 6 and 7. It is interesting to note that not all the cations can be substituted into III-V compounds.

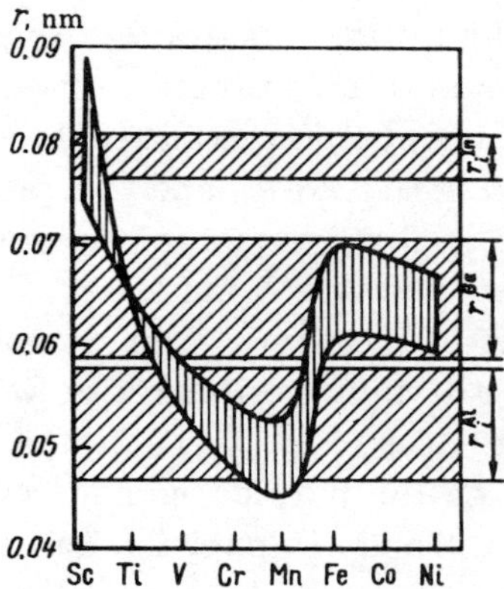

FIG. 6. Radii of transition metal ions and of cations in III-V semiconductors.

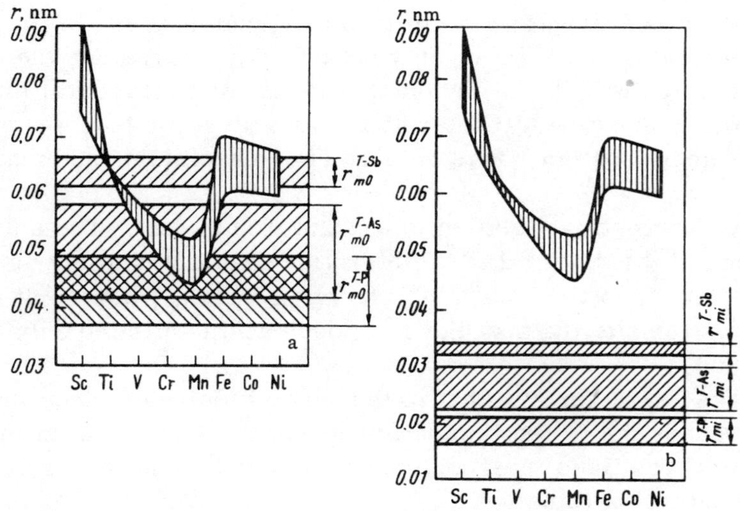

FIG. 7. Radii of transition metal ions and tetrahedral interstices with anion nonionized (a) and ionized (b) environments in III-V semiconductors.

We shall now consider in turn all III-V compounds. It follows from Figs. 6 and 7 that the ions of titanium, iron, cobalt, and nickel can replace gallium in GaAs and they cannot occupy tetrahedral interstices with the arsenic environment. On the other hand, the ions of vanadium, chromium, and manganese do not have to satisfy the cation substitutional condition $r \approx R$ and their dimensions agree with the dimensions of tetrahedral interstices with a nonionized arsenic environment. It is characteristic that the size factor makes the scandium ion unsuitable for the replacement of a gallium ion or occupation of an arsenic tetrahedral interstice. The Ga cations in gallium phosphide can be replaced with the same T-metal ions as in the case of GaAs, and the radius of the tetrahedral interstice with the nonionized anion environment in GaP matches reliably only the size of the Mn^{3+} ion.

In GaSb the ions of titanium, iron, cobalt, and nickel can occupy the Ga cation site and the interstice with the tetrahedral environment of antimony atoms.

In AlAs the Al cation may be replaced with vanadium, chromium, or manganese. These impurities can also occupy the As tetrahedral interstice. The dimensions of the other T-metal ions do not fit the dimensions of the cation or of the As tetrahedral interstice.

In InAs only the scandium ion satisfies dimensionally the condition for replacement of an indium ion. The same ions can occupy the As tetrahedral interstice in indium arsenide as in GaAs.

By analogy with InAs, only the Sc^{3+} ion can replace the indium cation in InSb, but – in contrast to the As tetrahedral interstice, the Sb tetrahedral interstice may accept the ions of iron, cobalt, nickel, but not those of vanadium, chromium, and manganese.

An interesting situation is encountered in InP. The small dimensions of the P tetrahedral interstice in this compound means that only the ions Cr^{3+} and Mn^{3+} can be accepted, whereas at the indium cation site the size factor is once again satisfied only by Sc^{3+}.

It is characteristic that in the case of ionized ligands in III-V compounds the T-ion impurities cannot be accepted at all by the interstitial positions irrespective of whether they have tetrahedral or octahedral coordinations.

In spite of the apparently convincing nature of the above analysis, it should be treated with great caution. The main reason for the scepticism is the well-known significant variation of the size of the T-metal atoms in compounds formed with elements in groups IV-VII.[2,3] The T-metal impurities in the semiconductors under consideration here do not form (in the literal sense) compounds with the atoms of the host semiconductor, but their behaviour is undoubtedly affected by the main features of the chemical binding appropriate to the relevant compound. The size of the T-metal atoms in compounds with anions is affected strongly by the number of the d-electrons, by the electron configuration, and by the symmetry of the coordination polyhedron formed by the surrounding anions.

This applies particularly to those T-metal ions which do not have a spherically symmetric ground S state and experience a strong nonisotropic influence by the anions located at the vertices of the coordination polyhedron. This results in stabilization of the energy of a crystal lattice with a T-ion at the centre of a coordination polyhedron by the crystal field and, as a consequence, in a considerable reduction of the T-ion size compared with that expected only in the presence of the Coulomb interactions when there is no crystal field.[2]

The reduction in the size of the T-ions, for example that observed in the case of the vanadium ion for an octahedral ligand symmetry,[4] can reach 25%. For the other T-ions the compression is somewhat less. It follows from general considerations that on transition to a tetrahedral coordination the compression of the T-ions by the crystal field should be even greater than in an octahedron because $y_T < y_O$.

Pearson[2] states that "any attempts to construct a general system of atomic radii of transition metals applicable to the compounds of these metals with group V-VII elements are undoubtedly doomed to fail." It remains not only to agree with this conclusion but to extend it to the compounds of the T-transition metals with group IV elements and to the ions and atoms of the T-transition metals present as impurities in IV, III-V, II-VI, and other semiconductors.

The evidence in Figs. 5-9 and an allowance for the compression effect shows that the conclusion about the positions of the T-ions ions in silicon and germanium interstices is strengthened still further. Moreover, they occupy most probably not the octahedral interstices, as would follow from a simple analysis, but the tetrahedral positions.

The realization of the T-ions in III-V compounds by the crystal field makes it even more difficult to replace the In cations, possibly makes easier the introduction of the T-ions (at least some of them) into tetrahedral interstices with nonionized P ligands, and clearly does not reduce the dimensions of T-ions sufficiently for them to occupy either tetrahedral or octahedral vacancies surrounded by ionized phosphorus, arsenic, or antimony ligands.

The qualitative conclusion of this crystallochemical analysis is that the preferred states of the 3d-metal impurities are those in the tetrahedral interstices of silicon and germanium and at the substitutional cation sites of gallium and aluminium in III-V compounds. The indium cation sites in III-V compounds cannot be occupied, because of the size considerations, by the 3d-metal atoms, except at interstitial positions with the cation environment.

The positions of the T-ions in the crystal lattices of semiconductors are governed not only by the size factor, but also by the nature of the chemical binding with the host atoms in a crystal. The approach used in such cases is based on the concept of the difference between the electronegativities $\Delta\chi$, but it gives the correct results only occasionally for a limited number of semiconductor–impurity systems. The attempts to apply it to the T-ions in semiconductors are pointless because of the large changes experienced by the electronegativity χ depending on the nature of the chemical compound in which the T-ion in question is located. Such indeterminacy in the selection of the values of χ applies to the atoms of many elements,[1] but it is manifested particularly strongly in the case of transition metal atoms. We may therefore conclude that there is as yet no crystallochemical theory capable of predicting "from first principles" the state of the T-metal impurities in various semiconductor crystals.

It is much easier to analyse theoretically the inverse problem: if we know the position of the T-ion in the lattice of a semiconductor, we can find its electron configuration and the charge state, which can then be checked experimentally.

It has been found that this task can be performed successfully by the crystal field theory. This theory is based on the equivalence of the spectroscopic elements describing electron terms and of the coordination of polyhedra (tetrahedra, octahedra, etc.) of which the crystal structure is composed. A polyhedron is regarded as consisting of negatively charged ligand ions located at its vertices and a positively charged T-metal ion located at the centre of a polyhedron.

The main characteristic of the crystal field theory is the neglect of the electron structure of the ligands, i.e. ligands are regarded as point electric charges. Their role reduces simply to the creation of an electric field, which is called the crystal field.

The crystal field symmetry is governed by the symmetry of a polyhedron which is a component of the structure of the crystal in question. It follows from Fig. 3 that in IV and III-V semiconductors we need consider only two types of symmetry: tetrahedral T_d and octahedral O_h. We shall demonstrate how the electron terms of the central T-ion are transformed in the crystal field of the ligands by considering first the spatial distribution of the electron structure of the T-ion in its free state.

The shape, dimensions, and distribution of the electron density in an atom are described by atomic orbitals (AOs) representing the wave eigenfunctions of electrons ψ_{nlm_l} which are solutions of the Schrödinger equation

$$H\psi_{nlm_l} = E\psi_{nlm_l} \tag{1}$$

and are governed by the following quantum numbers: principal n, orbital l, and magnetic orbital m_l. The wave function can be represented in terms of spherical coordinates as follows:

$$\psi_{nlm_l}(r, \vartheta, \varphi) = NR_{n_l}(r)Y_{lm_l}(\vartheta, \varphi), \tag{2}$$

where $R_{n_l}(r)$ is the radial part of the complete wave function ψ_{nlm_l} (r, ϑ, φ), which governs the dimensions and distribution of the electron density in AOs; $Y_{lm_l}(\vartheta, \varphi)$ is the angular component of the complete wave function, dependent only on the angles and governing therefore the shape of the AOs; N is the normalization factor showing that the probability of finding electron throughout the full range of variation of the functions R and Y should be equal to unity.

The radial part of the wave function for the d-electrons is of the form

$$R_{nl}(r) = (4/81)\sqrt{30}\,r^2 \exp(-r/3). \tag{3}$$

The function Y can itself be represented by a product of two functions $\theta_{lm_l}(\vartheta)$ and $\Phi_{lm_l}(\varphi)$, the values of which for the hydrogen atom are listed in Table III for the case when $m_l = l, l-1, \ldots, 0, \ldots -(l-1), -l$ for $l = 2$, i.e. for the d-electrons.

The magnetic quantum number m_l governs the orientation of the AOs relative to the Z axis. In the case of the d-electrons it assumes the values 2, 1, 0, −1, −2. We shall designate the corresponding AOs by d_2, d_1, d_0, d_{-1}, and d_{-2}.

It is clear from Table III that only d_0 is real and all the other AOs are complex functions. It is more convenient to use the functions in which the imaginary part is excluded, i.e. the functions

TABLE III. Angular Components of Atomic Orbitals (Wave Functions) of d Electrons (l = 2)

m_l	Designation	Complex part	$\theta_{lm_l}(\vartheta, \varphi)$	$Y_{lm_l}(\vartheta, \varphi)$	$Y_{lm_l}(\frac{x}{r}, \frac{y}{r}, \frac{z}{r})$
2	d_2	$\frac{1}{\sqrt{2\pi}}e^{2i\varphi}$	$-\frac{\sqrt{15}}{4}\sin^2\varphi$	$-\frac{1}{\sqrt{2\pi}}\sqrt{\frac{15}{16}}e^{2i\varphi}\sin^2\vartheta$	–
1	d_1	$\frac{1}{\sqrt{2\pi}}e^{i\varphi}$	$\frac{\sqrt{15}}{2}\sin\vartheta\cos\vartheta$	$\frac{1}{\sqrt{2\pi}}\sqrt{\frac{15}{4}}e^{i\varphi}\sin\vartheta\cos\vartheta$	–
0	$d_0 = d_{z^2}$	$\frac{1}{\sqrt{2\pi}}$	$\frac{\sqrt{5}}{2}(3\cos^2\vartheta - 1)$	$\frac{1}{\sqrt{2\pi}}\sqrt{\frac{5}{4}}(3\cos^2\vartheta - 1)$	$\frac{1}{\sqrt{\pi}}\frac{\sqrt{5}}{8}\frac{(3z^2 - r^2)}{r^2}$
–1	d_{-1}	$\frac{1}{\sqrt{2\pi}}e^{-i\varphi}$	$-\frac{\sqrt{15}}{2}\sin\vartheta\cos\vartheta$	$-\frac{1}{\sqrt{2\pi}}\sqrt{\frac{15}{4}}e^{-i\varphi}\sin\vartheta\cos\vartheta$	–
–2	d_{-2}	$\frac{1}{\sqrt{2\pi}}e^{-2i\varphi}$	$-\frac{\sqrt{15}}{4}\sin^2\vartheta$	$-\frac{1}{\sqrt{2\pi}}\sqrt{\frac{15}{4}}e^{-2i\varphi}\sin^2\vartheta$	–
1	$\frac{1}{i\sqrt{2}}(d_1 + d_{-1}) = d_{yz}$	–	–	$\frac{1}{\sqrt{\pi}}\sqrt{\frac{15}{4}}\cos\vartheta\sin\vartheta\sin\varphi$	$\frac{1}{\sqrt{\pi}}\sqrt{\frac{15}{4}}\frac{z}{r^2}$
–1	$\frac{1}{\sqrt{2}}(d_1 - d_{-1}) = d_{xz}$	–	–	$\frac{1}{\sqrt{\pi}}\sqrt{\frac{15}{4}}\cos\vartheta\sin\vartheta\cos\varphi$	$\frac{1}{\sqrt{\pi}}\sqrt{\frac{15}{4}}\frac{xz}{r^2}$
2	$-\frac{1}{\sqrt{2}}(d_2 + d_{-2}) = d_{x^2-y^2}$	–	–	$\frac{1}{\sqrt{\pi}}\sqrt{\frac{15}{16}}\sin^2\vartheta\cos 2\varphi$	$\frac{1}{\sqrt{\pi}}\sqrt{\frac{15}{16}} \times \frac{(x^2 - y^2)}{r^2}$
–2	$-\frac{1}{i\sqrt{2}}(d_2 - d_{-2}) = d_{xy}$	–	–	$\frac{1}{\sqrt{\pi}}\sqrt{\frac{15}{16}}\sin^2\vartheta\sin 2\varphi$	$\frac{1}{\sqrt{\pi}}\sqrt{\frac{15}{16}}\frac{xy}{r^2}$

which give the real images of AOs. These real wave functions of the d-electrons are obtained in the form of linear combinations of the functions d_2, d_1, d_{-1}, and d_{-2}, which are also listed in Table III.*

Since the T-ion should be located at the centre of a polyhedron and it is represented not by spherical but by rectangular coordinates, it is more convenient to convert Y also to the rectangular coordinates, which is done in the last column of Table III.

These coordinates are usually employed to identify the linear combinations of AOs by the lower indices. For example, we shall employ d_{z^2} (or, more exactly, d_{3z^2}), d_{xy} to denote the positions of an orbital along the Z axis and in the xy plane, respectively.

The positions of the d-orbitals, i.e. the angular dependence of ψ^2 in space, is shown graphically in Fig. 8. Four d-orbitals have the same eight-lobe shape and differ only in respect of the

*The signs of the linear combinations and of $\theta_{lm_l}(\vartheta)$ are selected so that $Y(\vartheta, \varphi)$ is positive. Details on the selection of the system of signs and a derivation of the function Y and its components θ and Φ can be found in L.D. Landau and E.M. Lifshitz's Quantum Mechanics: Non-Relativistic Theory (Theoretical Physics, Vol. 3), 3rd. ed., Pergamon Press, Oxford (1977), pp.89-92.

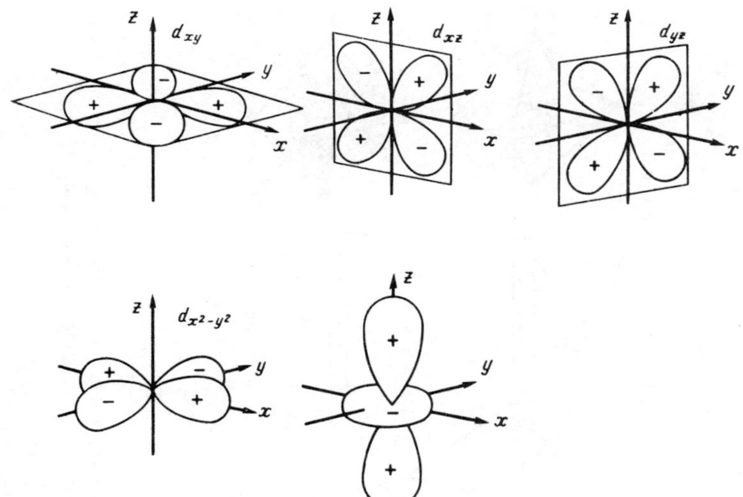

FIG. 8. Angular dependences of ψ^2 for d-electrons (shape of d-orbitals) in polar coordinates.

orientation relative to the coordinate axes. Only the d_{z^2} orbital is different. The reason for this is that there are influences of some AOs on others. In fact, the following distributions of d-orbitals are possible: d_{xy}, d_{xz}, d_{yz}, $d_{x^2-y^2}$, $d_{z^2-x^2}$, $d_{z^2-y^2}$. Out of the last three only two are independent. Therefore, a linear combination of any two can be used. It is usual to employ $d_{z^2-x^2}$ and $d_{z^2-y^2}$. A linear combination is described by $d_{z^2-r^2} = (1/\sqrt{2})(d_{z^2-x^2} + d_{z^2-y^2})$. The notation $z^2 - r^2$ is obtained using the expression $(z^2 - x^2) + (z^2 - y^2) = 2z^2 - (x^2 + y^2)$ and by the substitution $x^2 + y^2 = r^2$, which follows from the relationship between the rectangular and polar coordinates.

As long as the investigated T-atom is in its free state, all its AOs are equivalent and the electrons located on them have the same energy and they correspond to the same energy level. In this case it is usual to say that the d-orbitals of a free transition-metal atom are fivefold-degenerate.

We shall now place a T-ion with all its AOs into an octahedron and a tetrahedron (Fig. 9), and we shall select the coordinate axes x, y, z along the T-atom–ligand axes. The d-orbitals become inequivalent in an octahedron. The atomic orbitals d_{z^2} and $d_{x^2-y^2}$ are located in the direct vicinity of the ligands and their interaction with the latter (repulsion) is strong, whereas the other three orbitals d_{xy}, d_{xz}, and d_{yz} are located, respectively, between the axes x and y, x and z, y and z and are separated by large dist-

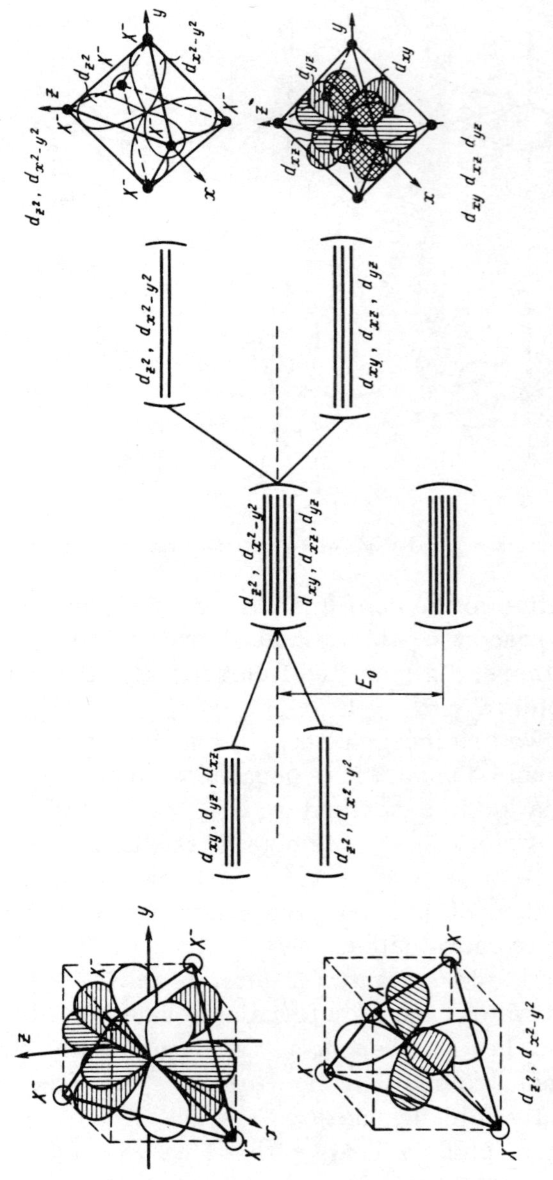

FIG. 9. Splitting of d orbitals of a T-ion in the crystal field. The situation in a tetrahedral field is shown on the left and that in an octahedral field is on the right.

ances from the ligands. Therefore, the electrons in these orbitals are affected less by the ligands. Consequently, there are two groups of electrons for different energies. The group of electrons on the orbitals d_{z^2} and $d_{x^2-y^2}$ has a higher value of the energy, whereas the group of electrons on the orbitals d_{xy}, d_{xz}, and d_{yz} has a lower energy. Therefore, a partial lifting of the fivefold degeneracy occurs in a crystal field with the O_h symmetry: triply and doubly degenerate states (levels) of the d-electrons now appear (Fig. 9).

In reality, the energy of an unsplit term of a T-atom located in a crystal is not equal to the energy of the term of a perfectly free T-atom. The energy difference E_0 (Fig. 9), represents the average energy of repulsion of the d-electron from the negative ligands on condition that the latter are distributed symmetrically around the central atom.

The reverse is true in a tetrahedron (Fig. 9): electrons on the orbitals d_{z^2} and $d_{x^2-y^2}$ are subject to a weaker electrostatic repulsion by the ligands because of the greater distance from the latter compared with electrons on the other (d_{xy}, d_{yz}, d_{xz}) orbials. Therefore, the splitting is reversed: a triply degenerate level (triplet) is now located higher, whereas a doubly degenerate level (doublet) is lower than the unsplit term of the T-atom. The two groups of the split orbitals are denoted by different symbols. Unfortunately, several different notation systems are encountered in the literature. The one-electron orbitals $d_{x^2-y^2}$ and d_{z^2} are denoted by d_γ, γ_3 (in accordance with Bethe notation) or by e_g (in accordance with the Mulliken notation), whereas the orbitals d_{xy}, d_{xz}, and d_{yz} are denoted by d_ϵ, γ_S, and t_{2g}, respectively.

All that we have said so far applies to the behaviour of the d-electrons in an atom but not to the atom as a whole, i.e. we have given the structure of an atom in the one-electron approximation based on the assumption that the electrons are independent.

The states of an atom allowing for the Coulomb interactions between electrons are described by the atomic quantum numbers L and S (orbital and spin numbers) and are known as terms. The quantum numbers of an atom are designated by Latin capitals and represent sums of the electron quantum numbers:

$$L = \Sigma l, \quad M_L = \Sigma m_l, \quad S = \Sigma S, \quad M_s = \sum_{2s+1} m_s$$

The terms are generally described by L, where L = S, P, D, F, G, H, I, depending on the value of the atomic orbital quantum number. In the case of the ground states of the atoms, the terms are deduced from the electron configurations. Let us consider the case of the d^n configuration. For the d-electrons we have $l = 2$ and $m_l = 2, 1, 0, -1, -2$, i.e. there are $(2l + 1) = 5$ values. Consequently, n electrons have to be distributed between five quantum cells. By way of example, we shall consider the case

TABLE IV. Valence States and Terms of T Ions in Tetrahedral Crystal Field

Free T ions		T ions in crystal field								
Ni	$3d^8 4s^2$							Ni^{4+}	Ni^{3+}	Ni^{2+}
Co	$3d^7 4s^2$							Co^{3+}	Co^{2+}	Co^{+}
Fe	$3d^6 4s^2$							Fe^{2+}	Fe^{+}	Fe^{0}
Mn	$3d^5 4s^2$	Mn^{7+}	Mn^{6+}	Mn^{5+}	Mn^{4+}	Mn^{3+}	Mn^{2+}	Mn^{+}	Mn^{0}	Mn^{-}
Cr	$3d^5 4s^1$	Cr^{6+}	Cr^{5+}	Cr^{4+}	Cr^{3+}	Cr^{2+}	Cr^{+}	Cr^{0}		
V	$3d^3 4s^2$	V^{5+}	V^{4+}	V^{3+}	V^{2+}	V^{+}	V^{0}			
Ti	$3d^2 4s^2$	Ti^{4+}	Ti^{3+}	Ti^{2+}	Ti^{+}	Ti^{0}				
Sc	$3d^1 4s^2$	Sc^{3+}	Sc^{2+}	Sc^{+}	Sc^{0}					
Electron configuration d^n		$3d^0$	$3d^1$	$3d^2$	$3d^3$	$3d^4$	$3d^5$	$3d^6$	$3d^7$	$3d^8$
Number of d^n electrons		0	1	2	3	4	5	6 (4)	7 (3)	8 (2)
Spin S = n/2		0	1/2	1	3/2	2	5/2	2	3/2	1
Term ^{2S+1}L		^1S	^2D	^3F	^4F	^5D	^6S	^5D	^4F	^3F
Splitting scheme of terms in tetrahedral field		^1S $^1A_1$	^2D: T_2, E	^3F: T_1, T_2, A_2	^4F: A_2, T_2, T_1	^5D: E, T_2	^6S $^6A_1$	^5D: T_2, E	^4F: T_1, T_2, A_2	^3F: A_2, T_2, T_1

when n = 3:

m = 2, 1, 0, −1, −2

↑	↑	↑		

Hence, we find that $M_L = \sum_3 m_l = 2 + 1 + 0 = 3$ and $S = \sum_3 S = (1/2) + (1/2) + (1/2) = 3/2$. Therefore, $L = M_L = 3$, i.e. we have the term M of multiplicity $2S + 1 = 4$ and finally we can see that the ground state of a T-atom with the d^3 configuration is ^4F.

The terms corresponding to the ground states of all the T-metal ions considered in the present monograph are listed in Table IV. When a T-ion is placed in a polyhedron with ligands, the terms split and the splitting is that shown in Fig. 9. The split terms are also denoted by capitals: Latin in accordance with the Mulliken system, for example $^2D \rightarrow T_{2g} + E_g$, or Greek in accordance with Bethe, for example $^2D \rightarrow \Gamma_5 + \Gamma_3$. The notation of the terms of the T-atoms in crystals is as follows:

Degree of accuracy	1	1	2	3	3	2	2	4
Notation:								
Bethe	Γ_1	Γ_2	Γ_3	Γ_4	Γ_5	Γ_6	Γ_7	Γ_8
Mulliken	A_1	A_2	E	T_1	T_2	$E_{1/2}$	$E_{5/2}$	G

The index g is used for a T-metal ion in a polyhedron with a centre of symmetry, for example, an octahedron. If this ion is located in a polyhedron which has no centre of symmetry (for example a tetrahedron), the index is omitted. We shall use the Mulliken notation.

Determination of the configuration and the magnitude of the splitting of the state of d^n atoms with different values of n in fields of different symmetry is the main quantitative task in the crystal field theory. This task is performed using group theory, but this topic is outside the scope of the present book. Group theory and its applications in an analysis of the term splitting has its own extensive literature (see, for example [5, 6]) and the reader is directed to this literature. We shall give only the main results of the crystal field theory needed for a better understanding of these applications to the behaviour of the T-ions in semiconductors.

The Hamiltonian of the T-ion in a crystal is of the form

$$H = \sum_{j=1}^{N} (P_j^2/2m - Ze^2/r_j) + H_{ee} + H_{cr} + H_{LS}, \tag{4}$$

where the first term is the sum of the number of electrons N and it represents the Hamiltonian of a free T-metal ion; H_{ee} is the electrostatic interaction (repulsion) of electrons in such a T-ion; H_{cr} is the interaction of the T-ion electrons with the crystal field; H_{LS} is the spin–orbit interaction.

The Hamiltonian (4) should include other weaker interactions of the spin–spin, spin–"foreign" orbit and other types, which become the dominant ones when details of spin resonance spectra (see Sec. 2 in the present chapter) and magnetic effects (see Sec. 1 in

Chap. 4) are interpreted.

In the case of ions with the Russell–Saunders coupling ($Z < 50$) we can distinguish the following three cases of the crystal field theory, depending on the relationship between the last three terms in Eq. (4):

1) a weak crystal field $H_{ee} > H_{LS} > H_{cr}$, which is usually encountered in the case of rare-earth ions with partly filled 4f- and 5f-shells;

2) a moderate crystal field $H_{ee} > H_{cr} > H_{LS}$, which is typical of transition metal ions from the iron group;

3) a stronger crystal field $H_{cr} > H_{ee} > H_{LS}$, which is found most frequently in the case of ions with partly filled 4d- and 5d-shells.

The Racah parameters A, B, and C and the parameters D_q and λ are used in the crystal field theory. The interactions described by these parameters are given in the table below.

Crystal field theory parameter	Type of interaction
$\Delta = 10D_q$	interaction of T-ion with crystal field H_{cr}; represents crystal field strength; governs splitting of T-ion terms in crystals
A	electron–electron interaction in T-ion; A = const for all d^n configurations
B, C	electron–electron interaction in T-ion (H_{ee}); separation of terms because of Coulomb repulsion of electrons
λ	spin–orbit interaction (H_{LS}); is a measure of interaction of orbit and spin moments of T-ion in crystals, resulting in additional splitting of T-ion levels

Since A = const, this parameter is usually ignored in the calculations. The quantity $\Delta = 10D_q$ is different for the same ion in fields of T_d and O_h symmetry: $D_{q\,tetr} = (4/9)D_{q\,oct}$.

The term splitting scheme in a tetrahedral field, deduced from a group-theoretic analysis, is given in Table IV.

We can demonstrate that the schemes for the d^n configurations in a tetrahedral field are identical with those for d^{10-n} configurations in an octahedral field. It follows from the crystal field

TABLE V. Electron Transitions Between Terms of T-Ions Split in Crystal Field of T_d Symmetry

Electron configuration of T-ion	Electron transition	Transition energy
d^1 (^2D)	$^2E \rightarrow {}^2T_2$	Δ
d^2 (^3F)	$^3A_2 \rightarrow {}^3T_2$	Δ
	$^3A_2 \rightarrow {}^3T_1$	$7.5\,B + 1.5\,\Delta - b^-$
	$^3A_2 \rightarrow {}^3T_1\ (^3P)$	$7.5\,B + 1.5\,\Delta + b^-$
	$^3T_1\ (^3F) \rightarrow {}^3T_1\ (^3P)$	$2b^-$
	$^4T_1 \rightarrow {}^4T_2$	$-7.5\,B + 0.5\,\Delta - b^+$
	$^4T_1 \rightarrow {}^4A_2$	$-7.5\,B + 1.5\,\Delta - b^+$
	$^4T_1 \rightarrow {}^4T_1\ (^4P)$	$2b^+$
	$^4T_2\ (^4F) \rightarrow {}^4A_2\ (^4F)$	Δ
d^4 (^5D)	$^2E \rightarrow {}^2T_2$	Δ
d^5 (^6S)	$^6A_1 \rightarrow {}^4T_1\ (^4G)$	$-\Delta + 10\,B + 6\,C - 26\,B^2/\Delta$
	$^6A_1 \rightarrow {}^4T_2\ (^4G)$	$-\Delta + 18\,B + 6\,C - 26\,B^2/\Delta$
d^6 (^5D)	$^5E \rightarrow {}^5T_2$	Δ
d^7 (^4F)	$^4A_2 \rightarrow {}^4T_2$	Δ
	$^4A_2 \rightarrow {}^4T_1$	$7.5\,B + 1.5\,\Delta - b^-$
	$^4A_2 \rightarrow {}^4T_1\ (^4P)$	$7.5\,B + 1.5\,\Delta + b^-$
	$^4T_1\ (^4F) \rightarrow {}^4T_1\ (^4P)$	$2b^-$
d^8 (^3F)	$^3T_1 \rightarrow {}^3T_2$	$-7.5\,B + 0.5\,\Delta + b^+$
	$^3T_1 \rightarrow {}^3A_2$	$-7.5\,B + 1.5\,\Delta + b^+$
	$^3T_1 \rightarrow {}^3T_1\ (^3P)$	$2b^+$
	$^3T_2\ (^3F) \rightarrow {}^3A_2\ (^3F)$	Δ
d^9 (^2D)	$^2T_2 \rightarrow {}^2E$	Δ

Note. The following notation is used in the above table: $b^+ = \frac{1}{2}[(9B + \Delta)^2 + 144B^2]^{\frac{1}{2}}$; $b^- = \frac{1}{2}[(9B - \Delta)^2 + 144B^2]^{\frac{1}{2}}$.

theory[6] that the splitting of terms of the d^1, d^4, d^6, and d^9 configurations is governed by just one parameter Δ (Table V). In the case of the d^2, d^3, d^7, and d^8 ions the splitting is governed by the parameters D_q, B, and C. In this case the expressions for the energy between the levels (if the field is of the T_d symmetry) are also included in Table V.

In the $3d^5$ configuration the term ^6S becomes $^6A_1$ (Table IV) and it is not split. Therefore, Δ, B, and C govern transitions between the ground state $^6A_1$ and the excited states formed from other terms, and are listed also in Table V.

A complete semiquantitative splitting scheme of the terms of ions with different d^n configurations in a tetrahedral field (or of d^{10-n} configurations in an octahedral field) is shown in Figs. 10 and 11, which represent the Tanabe–Sugano diagrams.[7] They are constructed by reducing even further the number of independent Racah parameters, because in many cases we have $B/C \approx 4$.

These diagrams represent the final result of the crystal field theory applied to the d^n configurations of T-metal ions. The diagrams are usually employed as the starting point in an analysis of

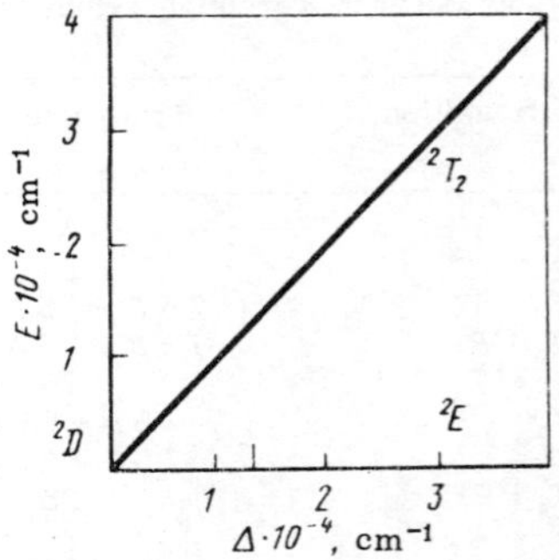

FIG. 10. Tanabe–Sugano diagram for energy levels of d^1 configuration in a tetrahedral field ($\Delta = 10D_q$).

the experimental results.

It is worth pointing out that these diagrams are of limited validity because of two circumstances: the diagrams are plotted only for one value of B and C and, moreover, the values of B and C are those of free T-ions. However, B and C for T-ions in crystals may be quite different. Moreover, these diagrams ignore the spin–orbit and other finer interactions.

All this makes such diagrams qualitative rather than quantitative. However, they do give an idea of the general behaviour of all the levels of a T-ion in a crystal field: they illustrate the relative positions of the levels and the relative distances between them. We shall now consider the general limitations of the crystal field theory which follow from the postulates assumed in developing this theory.

The most important is the representation of ligands as point structure-free charges which do not exchange electrons with the central T-ion. This model corresponds to the limiting case of 100% ionicity of the chemical binding. Therefore, for a long time the investigators have treated sceptically the use of the crystal field theory in the description of T-ions in chemical compounds or in IV and III-V crystals with a large proportion of covalence in the chemical binding. This sceptical attitude has been greatly undermined by the reports that the experimentally found parameters are in reasonable agreement with the crystal field theory "even in the case of such strongly covalent compounds as sulphide minerals."[6]

In the case of II–VI semiconductors it has been found that the parameters of the crystal field theory B and Δ obey the dependences[8,9]:

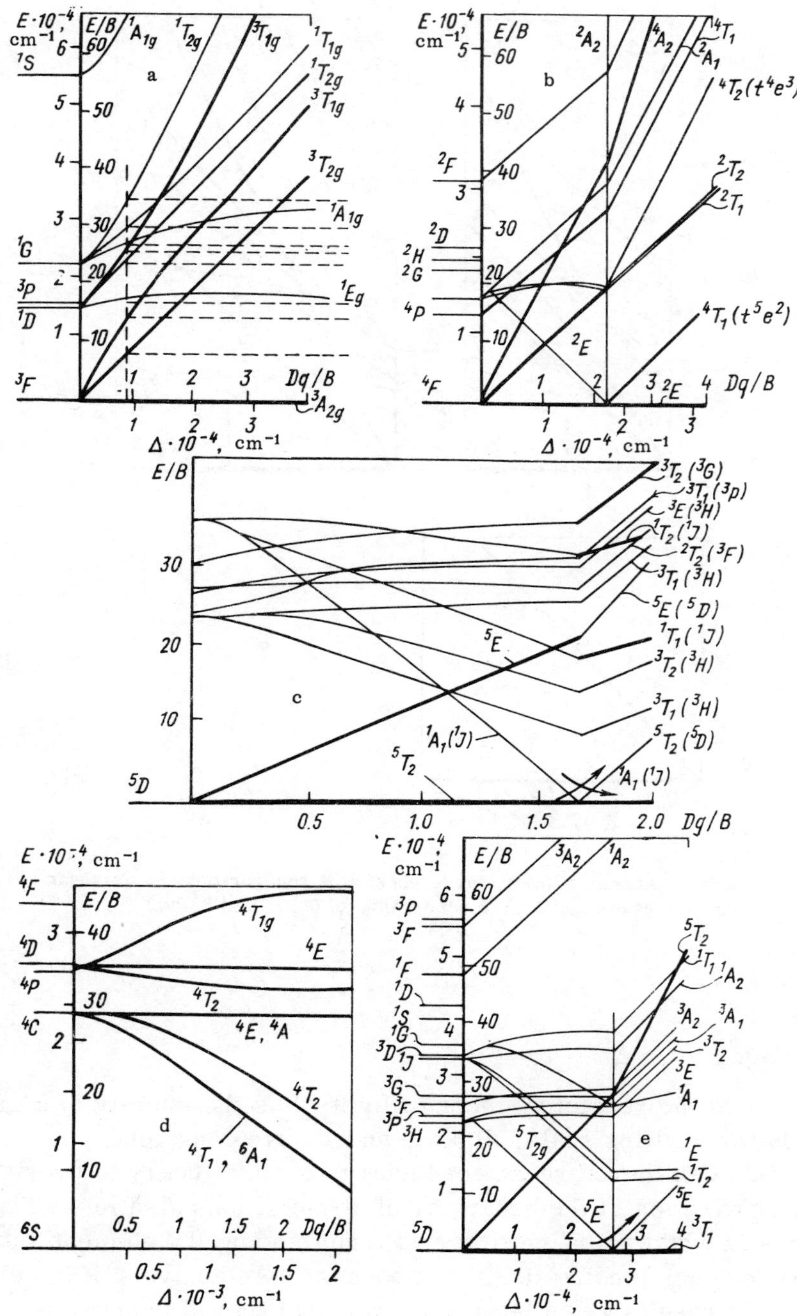
a
b
c
d
e
E/B
Dq/B
$\Delta \cdot 10^{-4}$, cm^{-1}
$\Delta \cdot 10^{-3}$, cm^{-1}

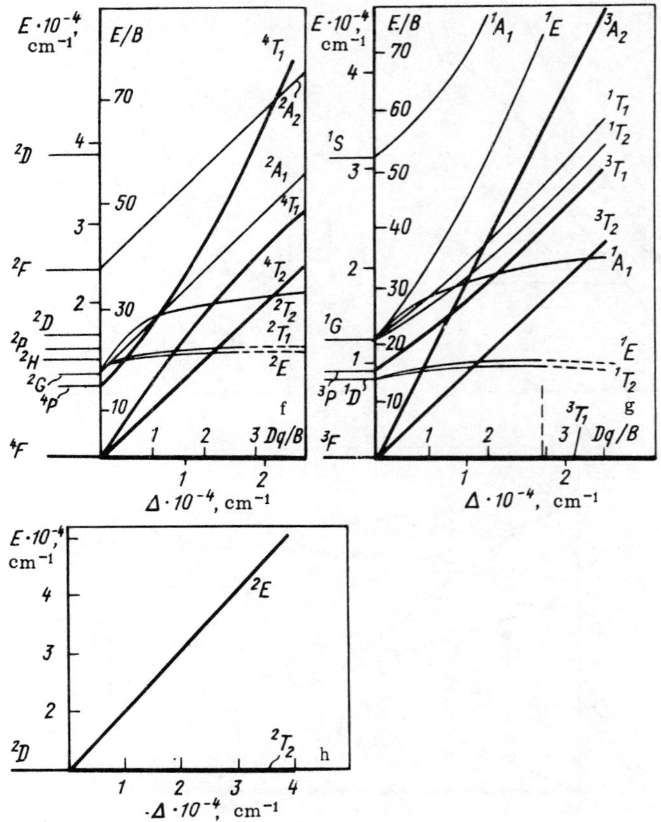

FIG. 11. Diagram of the energy levels of a d^n configuration in a tetrahedral field. Values of n: a) 2; b) 3; c) 4; d) 5; e) 6; f) 7; g) 8; h) 9.

$$B = B_0 n^2, \qquad \Delta = a_0 - b_0 f, \tag{5}$$

where n is the refractive index of light; f is the ionicity of a crystal on the Phillips scale; B_0, a_0, and b_0 are constants.

One can therefore expect the crystal field theory to provide an interpretation of the behaviour of T-metal ions also in semiconductors in which there is either no ionic binding (IV elements) or almost no such binding (III-V compounds). An analysis of the experimental results given later will show that this is indeed true. We may quote here the opinion of Moffitt and Ballhausen on this subject: "Much time will pass before another method will be developed to exceed the crystal field theory in respect of its simplicity, elegance, and force . . . . To our knowledge, no other molecular theory has given that many useful results and so close to the true situation, and none of them can be applied better directly in chemistry" (cited by Marfunin[5]). In fact, the crystal field theory has not become obsolete in its semi-empirical variant when the experimental results are interpreted qualitatively in accordance with the conclusions of this theory and the quantitative parameters are found experimentally.

A more rigorous theory is now available and it allows for the covalence of binding and for the electron structure of ligands. This is the ligand field or molecular orbital theory. It is being developed rapidly and applied in the descriptions of the chemical, optical, and magnetic properties of coordination compounds that contain T-atoms.[5,10] However, molecular orbital theory is cumbersome and it is very doubtful whether it should be applied to T-ions in semiconductors, since, strictly speaking, these ions do not form chemical bonds with ligands.

1.2. EXPERIMENTAL METHODS FOR INVESTIGATING STATES OF T-METAL IMPURITIES IN SEMICONDUCTORS

The most informative methods for the investigation of the state of 3d-metals in crystal lattices of semiconductors are those involving resonances, particularly electron spin resonance (ESR). This method is extremely sensitive and in modern apparatus the sensitivity can reach $\sim 10^{10}$ spin/G. It can be used to obtain information on the wave functions and spatial distribution of the electron density of paramagnetic ions and on the nature of the chemical binding in a crystal. Finally, ESR can be used successfully to determine the parameters of nonparamagnetic centres if their presence affects the ESR spectrum when various external agencies are applied.[11,12]

It is known that electrons have their own magnetic moment (spin). In the absence of an external magnetic field ($H = 0$) there can be two electrons with different spin directions in the same energy state.

The application of an external magnetic field $H \neq 0$ gives rise to two energy levels E_1 and E_2 in place of one (Zeeman splitting of levels). The upper level E_2 contains electrons with the spin oriented against the field ($S = -\frac{1}{2}$), whereas the lower one contains electrons with the spin oriented along the field ($S = \frac{1}{2}$). The Zeeman splitting of the levels $\Delta E = E_2 - E_1$ is proportional to the magnetic field intensity:

$$\Delta E = g\mu_B H, \quad (6)$$

where $\mu_B = e\hbar/4\pi mc$ is the Bohr magneton; g is the spectroscopic splitting factor given by the Landé formula

$$g = 1 + [J(J+1) + S(S+1) - L(L+1)]/2J(J+1), \quad (7)$$

where L and S are the quantum numbers of the orbital and spin momenta of an electron, respectively; J is the total (net) momentum along the field that acts on a particle.

Since any system tends to its minimum energy, under equilibrium conditions the number of electrons at the lower level is greater than at the upper level. The transfer of electrons to the upper level requires an additional energy amounting to ΔE. This energy can be provided by illuminating a crystal with light of energy $h\nu = \Delta E$.

If $h\nu$ = const, then the absorption of electromagnetic energy corresponding to the $E_1 \rightarrow E_2$ transitions occurs only if $h\nu = \Delta E = g\mu_B H$. The dependence of the absorption on the magnetic field exhibits a resonance. Such a single resonance line is obtained, for example, for free electrons.

Electron spin resonance can be observed not only for free electrons but also for atoms, but they must have a magnetic moment. The existence of this moment is governed by the electron shell structure.

In the majority of cases the valence electrons form filled shells and they are therefore diamagnetic. This excludes the possibility of observation of ESR of electrons belonging to atoms in the host lattice of IV and III-V semiconductors. Impurity atoms belonging to the iron group of metals, with charge states in which there is an odd number of electrons in the 3d-shell, should exhibit ESR. In the presence of an even number of electrons it is possible to observe ESR by altering the charge of an ion by some external agency (for example, illumination).

Electron spin resonance can be observed also when an impurity atom forms large complexes with unpaired spins as a result of interaction with vacancies or with other impurity atoms.

The resonance condition for a real crystal differs from that described by Eq. (6). This may be due to a variety of factors such as the spin–orbit interaction, admixture of excited states of an atom to the ground state, etc. One of the main reasons is the interaction between the magnetic moment and the nearest environment in the crystal lattice. The influence of this interaction can be linked formally to the change in the g-factor by Δg and by the addition of a field δH which increases the field acting on the spin from H to H + δH. Therefore, instead of Eq. (6), we now have $h\nu = \Delta E = (g + \Delta g)(H + \delta H)$. Since the internal field δH created by the crystalline medium and the corresponding resonance shift may be different for the various resonating atomic magnets, broadening or splitting of resonance lines may occur. Therefore, in a crystal each chemical element of those capable of exhibiting ESR has a specific spectrum consisting of a fixed number of lines and characterized by a certain distribution of positions.

The form of an ESR spectrum can be predicted using the spin Hamiltonian that is affected by the specific electron structure of a T-metal ion in a given crystallographic position and also by all the main interactions. If the T-metal ion is in an isolated state or in a cubic field, the spin Hamiltonian* is described by[13,14]:

$$H = g\beta\vec{H}\vec{S} + a/6[S_x^4 + S_y^4 + S_z^4 - - 1/5S(S+1)(3S^2 + 3S - 1)] + + A\vec{S}\vec{I} + \sum_n A_n\vec{S}\vec{I}_n - g_T\beta_T\vec{H}\vec{I}, \tag{8}$$

where S is the total spin of an electron; I is the spin of the nucleus of the T-metal atom; I_n is the spin of the ligand nuclei surrounding the T-metal ion; g_T is the spectroscopic splitting factor of the T-metal nucleus; β_T is the nuclear Bohr magneton; x, y, and z are the cubic axes of a crystal.

In the above expression the first term describes the usual Zeeman splitting in a magnetic field, whereas the second term represents the splitting in the crystal field (a is a constant). This term vanishes if $S \leq 3/2$. The third term reflects the magnetic interaction between the d-electrons of the T-ion impurity and its own nucleus, responsible for the hyperfine structure of the spectrum; A is the hyperfine interaction constant. The last term is included in the Hamiltonian when the spin of the T-ion nucleus of any T-metal isotope does not vanish.

The fourth term in Eq. (8) describes the superhyperfine interaction of the d-electrons not with the nucleus of the T-ion itself,

*In general, Eq. (8) is a polynomial of the total angular momentum J. If there is no orbital degeneracy and each of the $(2J + 1)$ levels described by the Hamiltonian (8) is nonmagnetic, then $J = S$. In the opposite case we have $J = L' + S$, where L' is the effective orbital angular momentum.

but with the nuclear moments of the ions in the environment which may consist of n different types of nuclei. Finally, the last term in Eq. (8) reflects the direct interaction of the T-impurity nuclei with the external magnetic field, which practically has no influence on the ESR spectra so that in describing these spectra the last term is ignored. However, it is essential in the case of electron-nuclear double resonance (ENDOR).

This double resonance appears when a material is subjected to electromagnetic fields with two different frequencies: one of these fields alters the electron quantum number and the other a nuclear quantum number. Changes in the nuclear subsystem then induce changes in the electron resonance.

Using Eq. (8), we obtain the following expression for the energy levels of an impurity atom:

$$E_{M,m} = Mg\beta H + AMm - g_T\beta_T H_m + \\ + \{M[I(I+1) - m^2] - \\ - m[S(S+1) - M^2]\} A^2/(2g\beta H_0). \quad (9)$$

This expression is considered in Ref. 13, but for the sake of simplicity the terms describing the splitting in a crystal field and the superhyperfine interaction with ligands are omitted.

In the case of a simple resonance we need consider only the transitions which alter the electron quantum number M (M_S in Fig. 12) by the amount $\Delta M = \pm 1$, whereas the nuclear quantum number m (m_I in Fig. 12) is unaffected ($\Delta m = 0$).

The following expression describes the relationship between the external magnetic field and the frequency of the observed electromagnetic radiation ν at which ESR occurs at spectral lines corresponding to transitions from a state (M − 1, m) to a state (M, m):

$$h\nu = g\beta H - Am + \\ + [I(I-1) - m^2 + m(2M-1)]A^2/2h\nu. \quad (10)$$

In double resonance experiments the selection rules governing the electron transitions are different: $\Delta M = 0$ and $\Delta m = \pm 1$. In this case, in place of Eq. (10), we find that the same transitions are described by[13]:

$$h\nu = |AM - g_T\beta_T H - [M(2m-1) + \\ + S(S+1) - M^2]A^2/2h\nu|. \quad (11)$$

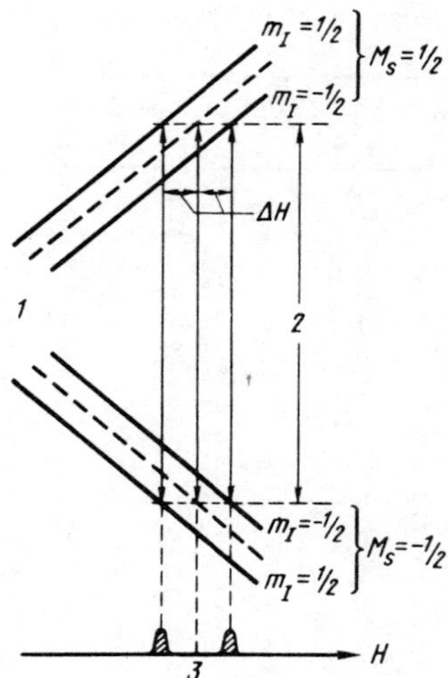

FIG. 12. Allowed transitions in ESR shown for the case when $J = \frac{1}{2}$ and $I = \frac{1}{2}$: 1) initial electron levels; 2) incident radiation photon $h\nu$; 3) line position in a field in the absence of the hyperfine splitting $h\nu/g\beta$.

The absolute value of the right-hand side of the above equation is considered because an electron in a state $(M, n - 1)$ may have a higher energy than in a state (M, m).

The nuclear quantum number m assumes any of the $(2I + 1)$ values in the interval from $-I$ to I. The electron quantum number can have any value $-S$ to S.

It therefore follows that an ESR spectrum should consist, in spite of lack of allowance for the interaction with ligands, of $(2I + 1)$ hyperfine interaction lines and each of these lines should split into 2S fine structure lines.

The line splitting due to the hyperfine interaction is governed, as indicated by Eq. (10), by the term denoted by δH:

$$\delta H = m(2M - 1)A^2/2h\nu. \quad (12)$$

The nature of the experimentally recorded ESR spectrum, i.e. the number of the hyperfine and fine structure lines, can be used to find I and S, respectively, so that the electron configuration and the charge state of the resonating ions can be determined. These quantities can be analyzed also a priori using the data in Tables IV and VI.

In Sec. 1 of the present chapter we have already mentioned the influence of ligands on the dimensions of T-metal ions in semiconductors. The influence of the crystal field must also be consid-

TABLE VI. Characteristics of Nuclei of T-Metals and of Component Elements of Semiconductors.

Isotope	Abundance, %	S*, h/2π units	Isotope	Abundance, %	S*, h/2π units
^{45}Sc	100	7/2	^{61}Ni	1.25	3/2
^{46}Ti	7.95	0	^{62}Ni	3.66	0
^{47}Ti	7.75	5/2	^{64}Ni	1.16	0
^{48}Ti	73.45	0	^{28}Si	92.27	0
^{49}Ti	5.51	7/2	^{29}Si	4.68	1/2
^{50}Ti	5.34	0	^{30}Si	3.05	0
^{50}V	0.24	0	^{70}Ge	20.55	0
^{51}V	99.76	7/2	^{72}Ge	27.37	0
^{50}Cr	4.31	0	^{73}Ge	7.67	9/2
^{52}Cr	83.76	0	^{74}Ge	36,74	0
^{53}Cr	9.55	3/2	^{76}Ge	7.67	0
^{54}Cr	2.38	0	^{27}Al	100	5/2
^{55}Mn	100	5/2	^{31}P	100	1/2
^{54}Fe	5.84	0	^{69}Ga	60.2	3/2
^{56}Fe	91.68	0	^{71}Ga	39.8	3/2
^{57}Fe	2.17	3/2	^{75}As	100	3/2
^{58}Fe	0.31	0	^{113}In	4.23	9/2
^{59}Co	100	7/2	^{115}In	95.77	9/2
^{58}Ni	67.76	0	^{121}Sb	57.25	5/2
^{60}Ni	21.16	0	^{123}Sb	42.75	7/2

*Nuclear spin.

ered when dealing with ESR spectra. The term in Eq. (8) describing this influence (containing the constant a) should be included for ions with $S > 3/2$.

In Chap. 2 we shall discuss in detail the influence of the crystal field on the energy spectrum of T-metal ions in semiconductors. Here we shall simply mention that the splitting of energy levels by the crystal field is manifested by the appearance of levels additional to those described by Eq. (9). Secondly, the interaction with the crystal field gives rise to an angular dependence of the ESR spectral lines. The positions of these lines can change on the H scale (for ν = const) proportionally to pa, where $p = (1-5)\varphi$; $\varphi = n_x^2 n_y^2 + n_y^2 n_z^2 + n_z^2 n_x^2$; n_x, n_y, and n_z are the direction cosines of the magnetic field relative to the cubic axes of a crystal.

The experimentally observed angular dependence of the ESR lines can be used to find the crystal field symmetry. In presenting the specific results for a series of T-impurity–semiconductor systems we shall demonstrate this procedure in greater detail.

It should be stressed that the tetrahedral symmetry of the interaction with the crystal field does not necessarily imply that the investigated T-metal ion is in a substitutional position (at a site) or in a tetrahedral interstice, because in both cases the T-metal ion is surrounded by four neighbours along the [111] axis of an f.c.c. semiconductor (Fig. 3). In contrast to ESR, the problem can be resolved at least in principle by an ENDOR spectrum, because these two positions are distinguished by the number and location of the second-nearest neighbours: a substitutional atom is surrounded by twelve second-nearest neighbours located along the [110] axis relative to the T-metal ion, whereas an interstitial atom has six second-nearest neighbours along the [100] axes.

The use of the ESR and ENDOR techniques in investigations of the state of the T-metal ions in semiconductors is restricted by two circumstances. First of all, only the T-ions with unpaired valence electrons, i.e. ions with an odd number of electrons in the d-shell, can resonate. This is due to the fact that these resonances result from the magnetic dipole coupling between a paramagnetic ion and an electromagnetic field. Moreover, in the case of a strong electron–phonon coupling of a T-ion in a semiconductor crystal the spectral lines become so broad that they are no longer observable.

In such cases it is possible to replace ESR successfully by the method of acoustic spin resonance (ASR). The ASR method is a complete analog of the ESR technique but it involves the absorption of hypersonic vibrations ($\nu > 10^9$ Hz). An ASR signal appears when the energy of acoustic quanta becomes equal to the energy of electron transitions between split levels. In ASR a strong electron–phonon coupling has a favourable influence on the intensity of the hypersonic absorption lines. Moreover, the ASR method can be used to study the state of T-ions with even or odd numbers of electrons in a partly filled d-shell.

Many impurities in semiconductors are in a state of association with other point defects. A special case of such associated defects is represented by impurity (ion) pairs. Impurities in the form of T-metal ions are not exceptional in this respect and once again it is useful to apply ESR mainly because the ESR spectrum of a pair consists of 2S fine-structure lines each of which splits into $(2I_1 + 1)(2I_2 + 1)$ hyperfine structure components, where I_1 and I_2 are the nuclear spins of the T-ion and of the second impurity in a pair. This circumstance makes it possible to identify the T-ions in the state of association with other accidental or specially introduced impurities. However, the large number of lines in an ESR spectrum makes it difficult to interpret the results. A further complication of this spectrum arises from the fact that different pairs may be oriented in different ways in a crystal relative to the crystallographic directions. The most typical situation, at least in silicon, is that when the orientation of the pair axis is along the [111] crystallographic direction, i.e. in the direction from a crystal lattice site to the nearest interstitial pair.

In this case the ESR spectrum of a pair can be described conveniently by a Hamiltonian "linked" to the pair axis if this axis is assumed to lie along an axis in a Cartesian coordinate system (x, y, z). Then, the Hamiltonian becomes[15]:

$$\begin{aligned} H = {} & \beta[g_{\|}S_zH_z + g_{\perp}(S_xH_x + S_yH_y)] + \\ & + DS_z^2 + a/6[S_\xi^4 + S_\eta^4 + S_\zeta^4 + \\ & + \frac{F}{180}[35S_z^4 - 30S(S+1)S_z^2 + 25S_z^2] + \\ & + \sum_{i=1}^{2}[A_iS_zI_{iz} + B(S_xI_{ix} + S_yI_{iy}) + \\ & + P_iI_{iz}^2 - (\gamma_i + R_{i\|})\beta_N I_{iz}H_z - \\ & - (\gamma_i + R_{i\perp})\beta_N(I_{ix}H_x + I_{iy}H_y)], \end{aligned} \tag{13}$$

where the cubic axes of a crystal are denoted by the indices ξ, η, and ζ. This Hamiltonian differs from Eq. (8) because it includes terms describing the interaction of the total spin S (angular momentum) and the nuclear spins I_i with a noncubic crystal potential and with an electric field gradient that appears in the presence of a pair. On the other hand, Eq. (13) is simplified by dropping the terms describing the superhyperfine interaction of the d-electrons of T-metal impurities with the nuclei of the host matrix and the interaction of the T-impurity nuclei with the external magnetic field, represented by the fourth and fifth terms in Eq. (8).

A solution of the Schrödinger equation obtained by the substitution of the Hamiltonian (13) depends on the relative contributions of the quantities occurring in Eq. (13). Various cases are considered in Refs. 15-18 and the reader is directed to this literature.

We shall consider here only two fundamentally different cases: $2D \approx h\nu$ and $|2D| \gg h\nu$.

In the former case the ESR spectrum consists of fine-structure lines due to the transitions characterized by $\Delta M = 2$ or 3 and $\Delta m = 0$, and of hyperfine structure lines ($\Delta M = \pm 1$, $\Delta m = \pm 1, 2$).[16]

In the latter case the large splitting even when $H = 0$ leaves only one transition between the $M = \pm\frac{1}{2}$ states and the spectrum consists of a single fine-structure line splitting into $(2I_1 + 1)(2I_2 + 1)$ hyperfine structure components. It is characteristic that the g-factor of the fine-structure line varies from $g_{\parallel}$ to $(S + \frac{1}{2})g_{\perp}$ when the magnetic field is rotated relative to the impurity pair axis,[15] which makes it easier to detect the T-ion in the associated state.

In addition to electron resonances, one should also mention nuclear γ resonance (NGR), also known as the Mössbauer effect, which is used to study the properties of T-metal impurities in semiconductors.

The essence of the Mössbauer effect is resonance fluorescence (emission, absorption, and scattering) of γ photons without recoil, i.e. without loss of part of the energy in a recoil of a nucleus which emits or absorbs a γ photon. The technique used in recording the γ-resonance spectra and a general description of the applications of this technique in solid-state physics and chemistry can found in reviews.[19, 20]

Resonance transitions in nuclei are extremely sensitive to the smallest deviations from the resonance energy. Therefore, the transfer of part of the energy of γ photons in nuclei, known as the recoil energy R and occurring in order to obey the law of conservation of momentum, "destroys" the resonance conditions. The value of R decreases on increase in the mass of the emitter or absorber nuclei: $R = E^2/2Mc^2$. However, in those cases when a nucleus emitting or absorbing a γ photon is fixed in a crystal lattice, it is possible for the recoil momentum to be transferred not to a single nucleus but to the whole lattice. Then, M represents the sum of the masses of all the nuclei in a crystal. In this case the recoil energy is infinitesimally small and it no longer prevents the attainment of resonance conditions. It is this circumstance that makes it possible to observe the Mössbauer spectra of crystals with lattices consisting of atoms of γ-ray emitters or absorbers. The situation is different when such atoms are in a lattice of "foreign" and light nonresonating atoms. In this case the NGR spectrum can be observed, as shown theoretically,[21] because of a shift of the Mössbauer atom relative to the nearest neighbours as a result of its interaction with optical phonons.

In considering the likely applications of NGR to the study of T-metal impurities in semiconductors we must bear in mind the exponential reduction in the probability of observation of this effect on the reduction in the mass of a T-metal atom. Moreover, because of a general reduction in the density of nuclear levels (at a given excitation energy) on reduction in the mass of a nucleus, it is necessary to exclude from the number of possible γ-resonating atoms all those with masses $M < M_{Fe}$ (Ref. 20). Clearly, there are restrictions also on the number of crystal matrices in which we can study the state of T-metal ions by the NGR method. The most suitable are the matrices with a high degree of ionicity, because optical vibrations play an important role in their phonon spectra and this increases the probability of observation of NGR. Therefore, for given iron and cobalt nuclei the probability of observation of NGR spectra in III-V compounds should increase along the phosphide–arsenide–antimonide series. In the case of IV crystals, characterized by an almost 100% covalent binding, it is possible to observe NGR in germanium because the mass of atoms surrounding a T-ion is large, but it is practically impossible to observe this resonance in silicon.

There are two variants of γ-resonance spectroscopy – emission and absorption – and the former is preferable for the investigation of T-metal ions. This is due to the fact that in the absorption case it is necessary to ensure that the concentration of the γ-absorber in a sample is high (10^{18} cm^{-3}). On the other hand, if the γ-ray emission spectrum is used, the sensitivity of the method rises so much that it is possible to observe T-metal ions acting as emitters in amounts of the order of 10^{15} cm^{-3}.

We must also bear in mind that when a Mössbauer T-atom is placed in a semiconductor lattice, the charge symmetry of its local environment may be disturbed and this is manifested by broadening of the NGR spectral lines or by their splitting due to the quadrupole interactions.

Finally, the emission variant of the NGR method has one other important feature. The state of, for example, iron in a crystal can be studied by introducing a "parent" nucleus of the cobalt isotope which as a result of radioactive decay is converted into an excited iron nucleus. In this case the positions of the detected iron atoms may be governed entirely by the lattice positions of the cobalt atoms introduced initially.

In spite of such serious restrictions, the applications of NGR have still provided valuable information on the state T-impurities in semiconductors. The main parameters of the NGR spectral lines are the chemical shift, i.e. the change in the energy position of a T-nucleus line depending on the crystal lattice in which it is located, and the quadrupole splitting of a line; they are both determined by the structure of the valence electron shells of the T-atom and by the nature of the interaction with the immediate ligand environment.

The simplest task is the separation of the covalent and ionic binding between the T-atom and the atoms of the host crystal. However, even this simple task is not easy without special calculations. If a quantitative interpretation of the NGR spectra is needed, it is necessary to calculate the electron density $|\psi(0)|^2$ and the electric field gradient $\partial^2V/\partial x^2$ in the region where a T-impurity atom is located. We must remember that separation of the binding into a ionic and covalent components represents a rough approximation which is far from adequate for a description of the fine features of the chemical binding that occurs when a T-metal impurity interacts with ligands in semiconductor crystals. In addition to resonance methods, useful information on the state of T-impurity atoms is provided by a study of the static magnetic susceptibility of semiconductors containing such impurities (the results of such studies will be considered in Chap. 4).

Finally, classical optical spectroscopy of T-metal ions in semiconductor crystals also plays a role. In the case of chemical compounds of transition metals with oxygen, sulphur, and halogens this method is quite well developed. It is the optical absorption spectra combined with the Tanabe–Sugano diagrams (see Figs. 10 and 11) that have largely made it possible to determine the Racah parameters for a large number of chemical compounds with T-metals[6] and thus confirm the validity of the crystal field method in describing the states of these metals.

The absorption bands appear in the optical spectra when when electrons undergo transitions from the ground state to one of the excited levels that are formed under the influence of the crystal field. For example, it is clear from Table IV that in the case of T-ions of the d^1 and d^6 configurations the absorption of optical photons should transfer electrons from the E levels to the T_2 levels, whereas in the case of the d^2, d^3, d^7 and d^8 configurations the spectrum has several absorption bands corresponding to electron transitions between three terms A_2, T_2, and T_1.

Optical transitions are subject to various selection rules, which allow or forbid some of them. For example, the selection rules require that changes in the orbital quantum number I should be restricted to ± 1 in electron transitions ("parity" selection), that the spin or multiplicity $(2S + 1)$ should be constant (spin selection), etc. Details of the selection rules are considered in Ref. 6.

In the case of real crystals there are various deviations from the selection rules associated with the mixing of d-states with others (for example, with p-states) and also associated with the spin–orbit interaction, the interaction of d-electrons with phonons, etc. Moreover, forbidden transitions still appear in the optical spectra, but in the form of low-intensity absorption bands which are one or two orders of magnitude weaker than the bands due to allowed transitions.

None of the currently used methods for investigating the electron structure of metals, including those considered in the present section, can give exhaustive information on the nature of the binding of T-metal impurities in semiconductors. Therefore, it is necessary to compare and supplement the results obtained by different methods on the same samples.

1.3. STATE OF T-METAL IONS IN SILICON AND GERMANIUM

The majority of the T-metal ions in the iron group have been investigated in silicon by the electron spin resonance (ESR) method. In the case of germanium the ESR method has been applied only to two ions: manganese and nickel. It should be pointed out that studies of the ESR in germanium are hindered mainly because of a strong spin–orbit interaction of the T-impurity electrons. This gives rise to relatively short spin–lattice relaxation times, which makes it difficult to observe ESR. Moreover, according to some authors,[15,22] an important circumstance is also the lower solubility of the T-metal ions in germanium compared with silicon. However, this circumstance should not play a significant role since both the residual (accidental) impurity background and the concentrations of defects in the crystal lattice are lower in the case of germanium. Finally, the sensitivity of modern ESR spectrometers is between two and three orders of magnitude higher than the concentrations of the T-metal impurities that are introduced into germanium by diffusion or during growth.

Scandium and Titanium

There are no published data on the state of these impurities in silicon and germanium. There are no known attempts to dope silicon or germanium with scandium or titanium with the aim of identifying their states in these elements, with the exception of the report by Fahrner and Goetzberger[23] who determined only the energy level of titanium in the band gap of silicon but none of the other characteristics relating to the nature of this level.

Vanadium

Woodbury and Ludwig[13] reported the results of an investigation of silicon samples containing vanadium in concentrations below 10^{15} cm^{-3} and an acceptor impurity forming shallow levels responsible for the p-type conduction of silicon. The ESR spectrum of vanadium ions has been observed only in such samples (Fig. 13).

In Chap. 3 we shall analyse in detail the energy spectra of T-metal impurities in IV and III-V semiconductors and we shall show that in most cases these impurities can exist in the form of ions with different charge states. The charge state can be controlled by double doping of a semiconductor. In this procedure a crystal is doped not only with a T-metal impurity, but also with atoms of a donor or acceptor impurity (concentration N_M) forming shallow energy levels in the band gap of the semiconductor and this shifts the Fermi level to the upper or lower half of the band gap. Consequently, the deep level of the T-metal ion is either occupied by electrons or it is vacant. Such double doping is a convenient procedure also because it suppresses the influence of the background of the accidental uncontrolled impurities N_d and N_a. This effect is achieved by ensuring that $N_M > N_d - N_a$ for an n-type crystal and $N_M > N_a - N_d$ for a p-type crystal.

In some cases this may give rise to impurity pairs, i.e. a chemical bond and not simple electrical compensation is established between the T-metal and shallow impurities. Formation of pairs affects the ESR spectrum and this will be considered in several examples below.

Since the Fermi level in silicon and germanium doped with vanadium is close to the top of the valence band, we may assume that the vanadium impurity is positively charged (if the energy level of vanadium lies above the Fermi level). The ESR spectrum (Fig. 13) consists of eight hyperfine structure lines, which gives $I = 7/2$. The spectrum shows clearly the splitting of each hyperfine structure line into three components, as shown on a larger scale in Fig. 14 for $m = \pm 7/2$. This splitting can be regarded as the fine structure with 2S lines. Hence, it follows that $S = 3/2$. The hyperfine interaction constant A can be found from Eq. (12).

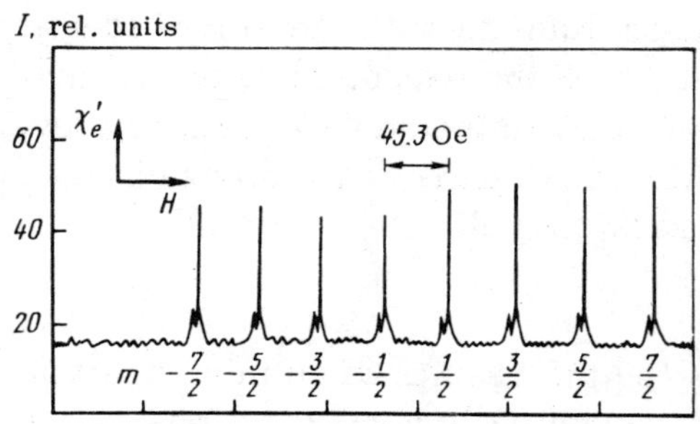

FIG. 13. Electron spin resonance spectrum of the V^{2+} ion in p-type silicon at 1.3 K (Ref. 13). Here, I is the line intensity.

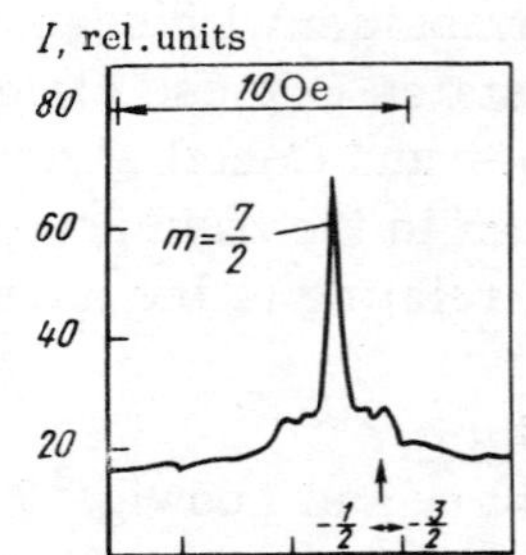

FIG. 14. Fine structure of the m = ±7/2 line in the ESR spectrum of the V^{2+} ion in silicon at 1.3 K (Ref. 13).

However, the fine structure lines are obviously insufficiently resolved. Moreover, according to Woodbury and Ludwig,[13] it is not clear why small peaks appear on both sides of the line maximum in Fig. 14. Therefore, they determined[13] the spectrum of electron–nuclear double resonance (ENDOR) in order to find accurately the values of A and S. All the lines corresponding to $M = \pm\frac{1}{2}$ were detected. The ENDOR frequencies of the ^{51}V isotope were calculated for the $(m = \frac{1}{2}) \to (m = -\frac{1}{2})$ transitions using Eq. (11) and this was done for various values of S. In the experiments the ENDOR frequency of these transitions was 63.09 ± 7.68 MHz, which agreed with the frequency expected for S = 3/2.

This value of S in combination with the positive sign of the charge of the resonating ion provides grounds for assuming (Table IV) that the vanadium impurity in these experiments is present in silicon in the form of the V^{2+} ion and its electron structure is $3d^3 4S^0$. The values of A and of the g-factor taken from Ref. 13 are listed in Table VII.

It is characteristic that both the g-factor and the hyperfine interaction constant are isotropic and this means that the V^{2+} ion is in one of the positions with an undisturbed tetrahedral symmetry: a site or a tetrahedral interstice.

The vanadium ion is more likely to occupy the interstitial position in a tetrahedral void because the ENDOR spectrum is well accounted for by the values of S and A found from the ESR data and by the known g-factor of the vanadium nucleus in the free state. The constancy of the g-factor is evidence of the absence of

TABLE VII. Parameters of T-Metal Ions in Silicon Obtained from ESR Spectra

Ion	Configuration	S	J	g	A ($\times 10^4$, cm^{-1})	a ($\times 10^4$, cm^{-1})
			in tetrahedral interstices			
$^{51}V^{2+}$	$3d^3$	3/2	3/2	1.9892	−42.10	—
$^{53}Cr^{+}$	$3d^5$	5/2	5/2	1.9978	+10.67	+30.16
$^{53}Cr^{0}$	$3d^6$	2	1	2.97	15.9	—
		2	2	1.72		
$^{55}Mn^{+}$	$3d^6$	2	1	3.01	73.8	
		2	2	1.68	46.1	
		2	3	1.34	—	
$^{55}Mn^{0}$	$3d^7$	3/2	1/2	3.362	92.5	—
		3/2	3/2	1.46		
$^{55}Mn^{-}$	$3d^8$	1	1	2.0104	−71.28	—
$^{55}Mn^{2+}$	$3d^5$	5/2	5/2	2.0066	−53.47	+19.88
$^{57}Fe^{+}$	$3d^7$	3/2	1/2	3.524	2.99	—
$^{57}Fe^{0}$	$3d^8$	1	1	2.0699	6.98	—
$^{61}Ni^{3+}$	$3d^9$	1/2	1/2	2.026	3.6	—
			at regular sites			
$^{53}Cr^{0}$	$3d^2$	1	1	1.9962	12.54	—
$^{55}Mn^{+}$	$3d^2$	1	1	2.0259	−63.09	—
$^{55}Mn^{2-}$	$3d^5$	5/2	5/2	2.0058	−40.5	+26.1
$^{61}Ni^{3+}$	$3d^9$	1/2	—	[110] 2.0357	—	—
				[111] 2.0133		
				[112] 2.0333		

any stable chemical binding with ligands and this is more likely to be the situation when the vanadium ion occupies a tetrahedral interstice than in the case when it replaces a silicon atom. Unfortunately, no attempts have been made to supersaturate Si:V crystals with vacancies. In this case we would expect the capture of vacancies by the interstitial V^{2+} ions and their transfer to the regular sites in the crystal lattice.

Chromium

The ESR spectra of chromium impurities in silicon have been investigated quite thoroughly and interpreted.[13] It should be pointed out immediately that a careful search for a resonance of the chromium atoms in n-type silicon was unsuccessful. The ESR spectrum was observed only when chromium was introduced by diffusion into silicon samples already containing an acceptor impurity forming shallow energy levels. Hence, it follows that the observed ESR spectrum[13] is either due to neutral or positively charged chromium ions. The spectrum (Fig. 15) consists of five lines which were interpreted by Woodbury and Ludwig as the fine structure.

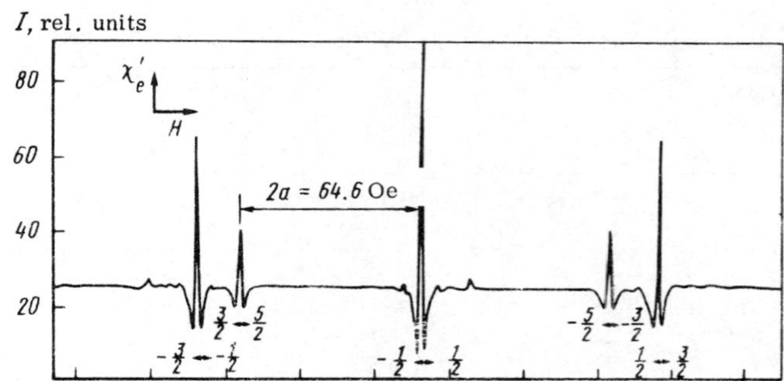

FIG. 15. Spectrum of the Cr^+ ion in silicon at 20.4 K (H ∥ [001]) (Ref. 13).

Then, equating the number of these lines to 2S, we find that S = 5/2, which corresponds (Table IV) to the Cr^+ ion. The experimental values of g and a found experimentally from this spectrum are 1.9978 and $30.16 \cdot 10^{-4}$ cm^{-1}. For the spin value 5/2 the Cr^+ ion is subject to the crystal field and, consequently, these fine structure lines should be anisotropic relative to the orientation of a crystal in a magnetic field. In fact, an angular dependence was observed for the ESR spectral lines (Fig. 16). The continuous curves in Fig. 16 represent the results of a theoretical calculation carried out substituting the experimental values of the g-factor and of the constant representing splitting by a cubic field. We can see that the experimental points agree well with the calculations.

Near each line in the ESR spectrum (Fig. 15) there are weaker lines due to the hyperfine interaction of the d-electrons of chromium with nuclei of the ^{53}Cr isotope. There is no doubt about the origin of these lines, because Woodbury and Ludwig[13] repeated the experiments introducing chromium enriched with the ^{53}Cr isotope to 96% into silicon by diffusion. The ESR spectrum obtained for such samples (Fig. 17) shows clearly that the intensities of the hyperfine structure lines are now much stronger. The identification of these lines as due to the hyperfine structure is supported also by the number of these lines, which is four, as expected because the nuclear spin of ^{53}Cr is 3/2 (Table VI). These results leave no doubt that the resonating T-impurity ions in these samples are indeed the chromium ions. Experiments show that the hyperfine interaction lines are isotropic, i.e. the Cr^+ ion is in a tetrahedral environment.

The ENDOR spectrum of samples with high concentrations of

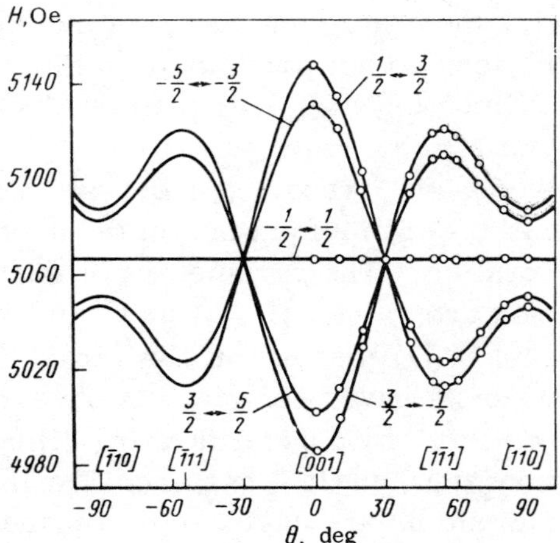

FIG. 16. Angular dependences of the ESR lines of the Cr^+ ion in silicon at ν = 14 170 MHz. The continuous curves are calculated[13] for the splitting in a crystal field of cubic symmetry using the experimental values of the g-factor and a. Here, θ is the angle between H and the crystal axis.

the ^{53}Cr isotope was determined at very low temperatures.[13] The resonance was observed for the M = $\pm\frac{1}{2}$, ±3/2, and −5/2 levels. The absence of ENDOR in the case of the m = +5/2 levels was explained by an extremely low population of these levels at the temperature of this experiment, which was 1.3 K.

Since these ENDOR frequencies corresponded to the frequencies calculated from Eq. (11) using the nuclear moment of

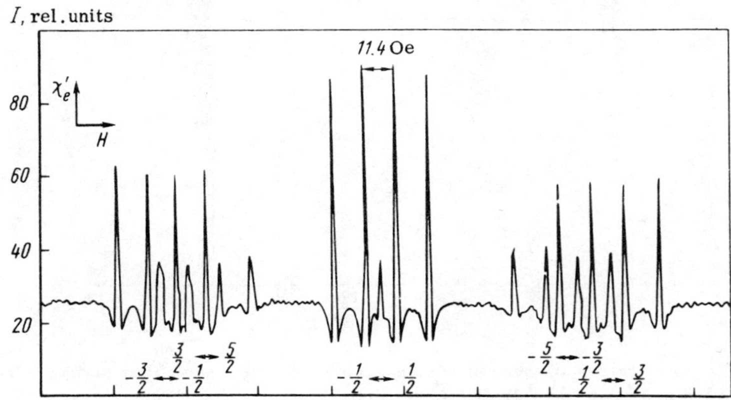

FIG. 17. Electron spin resonance spectrum of the Cr^+ ion enriched with the ^{53}Cr isotope, recorded for silicon at T = 20.4 K (Ref. 13).

chromium, known for the chromium nuclei in the free state, it follows that the Cr^+ ions do not participate in the formation of valence bonds in the silicon crystal, i.e. they are located in tetrahedral interstices of the lattice.

Ludwig and Woodbury[15] reported observation of the ESR spectrum of interstitial chromium atoms in the neutral state Cr^0 with the $3d^6$ configuration. The parameters of this spectrum are listed in Table VII for two values of J (1 and 2). The spectrum for one of these values (J = 1) is shown in Fig. 18.

Interesting attempts have been made to observe chromium in the substitutional position[24] by supersaturating silicon crystals with vacancies. Woodbury and Ludwig[24] assumed that in this case the interstitial chromium atoms are captured by vacancies and transferred to the substitutional positions. Such a redistribution was induced by diffusion saturation of a silicon sample simultaneously with two impurities: chromium and copper.[24] At the diffusion temperature the atoms of copper occupy mainly substitutional positions and during slow cooling the retrograde nature of the solubility results in the displacement of the excess copper from the regular sites so that the crystal becomes saturated with vacancies. The rate of migration of vacancies is high and even at low temperatures

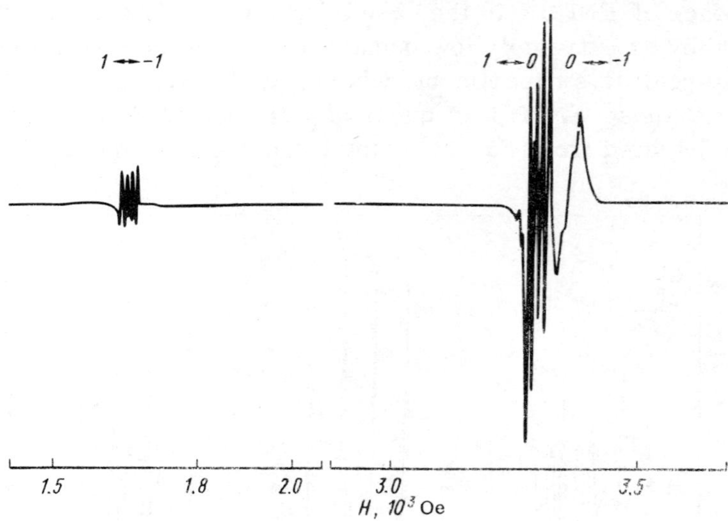

FIG. 18. Electron spin resonance spectrum of the Cr^0 impurity in an interstitial position in silicon (J = 1; H ∥ [001]) (Ref. 13).

there is a good probability that a vacancy encounters an interstitial chromium ion. This process can be described by the following quasi-chemical reactions:

$$Cu_s^- \rightarrow Cu_i^+ + V^- + e; \qquad V^- + Cr_i^+ \rightarrow Cr_s^0. \tag{14}$$

It is clear from the second reaction that an increase in the vacancy concentration produced by some metal or another should also result in a redistribution of chromium atoms from the interstitial to the site positions. In fact, Woodbury and Ludwig[24] observed the same type of the ESR spectrum of chromium as found after irradiation of copper-free Si:Cr crystals by electrons of energy 1.5 MeV. The ESR spectrum obtained in this case (Fig. 19) is given in Ref. 15. It consists of four lines of a hyperfine structure due to the interaction with nuclei of the ^{53}Cr isotope. Near each of these lines there are weaker lines due to a superhyperfine interaction with the ^{29}Si nuclei. The value of the g-factor is found to be 1.9962 and the hyperfine interaction constant is $A = 12.54 \cdot 10^{-4}$ cm^{-1}. The ENDOR measurements indicate that the electron spin is $S = 1$ and this, as demonstrated in Table IV, corresponds to the electron configuration $3d^2 4S^0$. Hence, we can draw the conclusion that a resonating chromium atom is located at a crystal lattice site. Four of its electrons (one from the 4S shell and three from the 3d shell) are used to form valence bonds in place of the electrons of the silicon atom originally present at the site. Consequently, two electrons remain in the d-shell of the

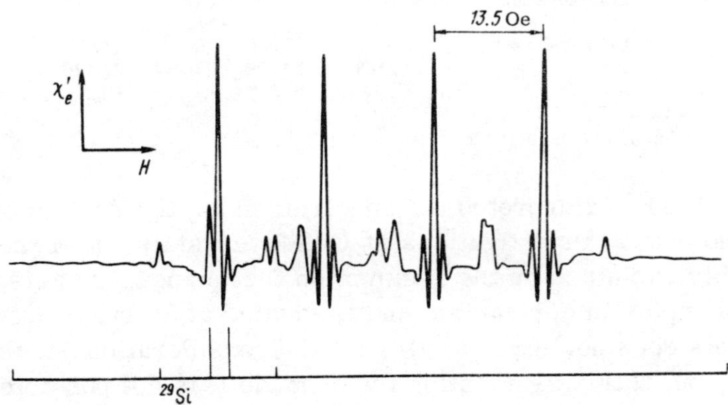

FIG. 19. Electron spin resonance spectrum of the $(^{53}Cr)^0$ impurity in a substitutional state in silicon (H ∥ [001]) (Ref. 15).

chromium atom. Such a state of a site chromium atom in silicon corresponds to Cr^{4+}, i.e. to the neutral state of Cr^0 relative to the silicon lattice. If the behaviour of the T-ions in silicon is considered from the crystal chemistry point of view (see Fig.5), it is found that the experimentally observed site state Cr^0 is in conflict with the size factor. Moreover, the even number of electrons in the d-shell of a site Cr^0 ion does not admit the appearance of ESR.

This contradiction is most likely due to the fact that the state of chromium does not represent simple substitution, similar to the substitution of silicon with shallow impurities, such as group III and V atoms.

We can assume formation of a complex associated defect $(V^{-}-Cr_i^{+}) = Z^0$, the geometry of which is more likely to be a manifestation of the Jahn–Teller effect, which may give rise to ESR even in the case when the number of electrons in the d-shell is even (see Sec. 1.2). The ESR spectrum should then be regarded as due to an impurity pair, as discussed below.

Manganese

The state of manganese impurities in germanium was investigated by the ESR method[25] but the spectrum was not reported. The parameters of the Hamiltonian describing this spectrum are as follows[25]:

Ion	$^{55}Mn^{2+}$	$^{61}Ni^{3+}$ (site substitution)		
Configuration	$3d^5$		$3d^9$	
S	5/2		1/2	
Principal axes	—	$[\bar{1}10]$	[001]	[110]
g	2.006	2.1128	2.0294	2.0176
$A \cdot 10^4$	–42.7*	10.3	≤ 1.6	12.2

*Value of a is $8.9 \cdot 10^{-4}$ cm^{-1}.

Watkins[25] interpreted the spectrum using the results of electrical and optical investigations of Ge:Mn crystals and these were reasonably explained on the assumption that manganese acts as a double acceptor in germanium and is located at a crystal lattice site. This does not agree at all with the considerations of the size (Fig. 5), but if this assumption is not made it is not possible to explain any of the electrophysical properties of Ge:Mn.

The site position of manganese in germanium and its double charge in the Mn^{2+} state corresponds to the $3d^5$ configuration so that, as expected, we have $S = 5/2$ and $g = 2$.

The introduction of manganese impurities into silicon produces a greater variety of ESR spectra, compared with T-metal impurities considered earlier. This is due to the fact that manganese atoms can occupy various crystallographic positions in silicon and can have various charge states.

In contrast to vanadium and chromium, the ESR signal of manganese is observed not only for p-type but also for n-type silicon. For example, it is reported in Ref. 13 that manganese was introduced by diffusion into silicon containing phosphorus. After diffusion at 1250° C the samples placed in a quartz ampoule were quenched rapidly by dropping the ampoule in water. This increased the concentration of ionized phosphorus donors P^+ by an amount approximately equal to the concentration of the manganese atoms introduced at the diffusion temperature. This observation alone was sufficient to deduce that manganese in these samples was in the form of a singly charged negative Mn^- ion. The ESR spectrum obtained for these crystals[13] is shown in Fig. 20. Preliminary experiments established that some of the lines in the spectrum belong to the ^{31}P isotope of phosphorus and to iron impurity atoms unavoidably captured together with manganese which was clearly not sufficiently pure. These "stray" lines are denoted by arrows in Fig. 20. The number of lines due to manganese is six, as expected for the hyperfine structure of manganese (Table VI). It is clear from Fig. 20 that each hyperfine structure line is split into two. If this splitting is attributed to the fine structure, a resonating atom should have an electron spin of unity and this corresponds (by reference to Table IV) to a Mn^- ion with the electron configuration $3d^8 4s^0$ if it is located at an interstice or to a Mn^{5+} ion at a substitutional position, but with the $3d^2$ configuration. The latter is clearly not realistic because there would have been no correlation between the concentrations of the manganese dopant and the increase in the amount of ionized phosphorus.

The value $S = 1$ was confirmed also by the ENDOR spectra of the $M = 0$ and $M = \pm 1$ levels.[13] The same spectra were used to determine the hyperfine interaction constant A, the value of which is given in Table VII. When Si:Mn samples were prepared by the same diffusion method followed by quenching but the starting material was in the form of crystals doped previously with a shallow acceptor (boron, gallium), the ESR spectrum was quite different. It was described in Ref. 13. It consisted of five fine-structure lines, each of which was split into six components by the hyperfine interaction. A comparison of the number of lines with the data in

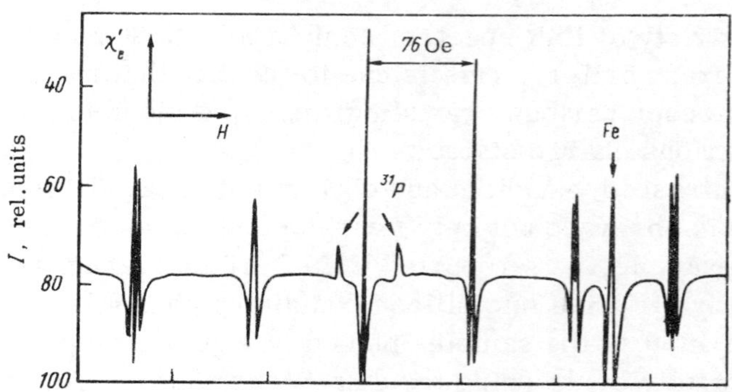

FIG. 20. Electron spin resonance spectrum of Si:Mn (H ∥ [111]; T = 20.4 K) (Ref. 13).

Table IV shows that in this case (S = 5/2) a resonance is observed for the Mn^{2+} ion with the electron configuration $3d^5$. The Mn^{2+} ion is subject to the crystal field. The splitting constant a and the other characteristics of the ESR spectrum of the Mn^{2+} ions in silicon are given in Table VII.

Woodbury and Ludwig[13] drew attention to the fact that the intensities of the lines in the spectra of Mn^- and Mn^{2+} were approximately the same (for the same doping conditions). Therefore, they concluded that both spectra represented the same crystallographic position of a manganese ion in a tetrahedral interstice, but in different charge states. However, it should be pointed out that the final proof has not yet been obtained because determination of the concentrations of the paramagnetic centres from the spectral line intensities is generally incorrect.

Strictly speaking, the number of paramagnetic centres in a sample is proportional to the area under the curve describing the absorption of microwave power: $N = K\int_{-\infty}^{\infty} U_{abs}(H)dH$, where the integral represents the area under the absorption curve and the proportionality coefficient K is given by

$$K = TVK_1/[\omega g S(S + 1)V_0 V_1 \eta Q_0], \tag{15}$$

where T is the absolute temperature; V is the volume of the sample; η is the resonator fill factor (the degree to which the resonator is filled by the sample); ω is the frequency at which the ESR is observed; Q_0 is the Q-factor of the resonator; V_0 and V_1 are quantities related to the frequency and power of the klystron supplying the microwave power; K_1 is a coefficient which contains universal constants.

The value of K depends on the experimental conditions and on the g-factor. Though, in principle, it is possible to keep constant the experimental conditions in the determination of the ESR spectra of different ions, the g-factors are different and this gives rise to different absorption curves. It is indeed possible to determine the concentration of centres from the areas under the lines, but a rigorous allowance for all the parameters has to be made in Eq. (15). However, there are only a few situations in which this can be done in respect of the line intensities.

Observation of the ESR spectrum of interstitial manganese ions carrying a single positive charge (Mn^+) was reported.[26] The characteristics of this spectrum are given in Table VII for three values of J.

An attempt has been made to force manganese ions to the site positions by supersaturating a silicon sample with vacancies.[24] This was achieved in the same way as in the experiments with chromium and it was the result of double doping of silicon with manganese and copper. When supersaturation of the sample with copper was relieved, the concentration of vacancies was high and, in accordance with the suggestion made in Ref. 24, these vacancies should capture interstitial manganese ions.

The initial samples doped simultaneously with manganese and copper already contained either phosphorus or boron. Consequently, it was possible to prepare two groups of samples. In one of them the excess of phosphorus (compared with copper and manganese) predominated, whereas in the other there was an excess of boron. In the former group of n-type samples the Fermi level was located in the band gap at a high position directly under the bottom of the conduction band. The manganese ions could therefore assume only a negative charge. In the absence of the interaction with vacancies the spectrum of the ion would have been the same as in the absence of copper (Fig. 20). However, the experiments revealed a completely different ESR spectrum. The value of the spin obtained in Ref. 24 was 5/2. We recall that in the case of n-type samples doped with chromium the ESR spectrum of the chromium ions was not observed. In samples of the second group the Fermi level was located at the bottom of the band gap and, therefore, a manganese ion should acquire a positive charge. The ESR spectrum, which in this case was similar to the spectrum of Cr^0 (Fig. 18), yielded S = 1. It was not clear whether the manganese ions were in this case singly or multiply charged. In Ref. 24 the authors began from the assumption that pure high-resistivity silicon shows no ESR spectrum on introduction of manganese. This means that the difference between the charges of manganese ions in both groups of samples (with phosphorus or boron) should be at least 2. However, the experimental results indicated that in the former case the total spin was governed by five electrons, whereas in the latter it was governed by two electrons. Consequently, the difference between the ion charges should be 2 and not 3. The smallest charge numbers corresponding to this condition are 1 and −2. Therefore, the manganese impurities in the first group of samples had the charge state Mn^{2+}, whereas those in the second had Mn^{5+}.

The parameters of the ESR spectrum of these ions are listed in Table VII. The charge numbers together with the values of the spin found from the spectra correspond (as is indicated by the data in Table IV) to the electron configurations $3d^5$ and $3d^2$ for the ion positions at the crystal lattice sites. As in the case of chromium, this conclusion is in conflict with the results of a crystallochemical analysis. However, the conclusion on the site positions of the Mn^{5+} and Mn^{2+} ions in silicon with a high concentration of vacancies is supported also by the higher time stability of the spectra of these ions, compared with the stability of the spectra of Mn^{2+} and Mn^- reported in Ref. 24. However, we cannot exclude the possibility that the spectra of Mn^{5+} and Mn^{2+} represent products of association of interstitial manganese ions with vacancies and not simple substitution at the sites.

Iron

The ESR spectrum of iron in silicon was analysed in Refs. 13, 27, and 28. The spectrum was investigated most thoroughly for compensated samples containing phosphorus in addition to the iron impurity introduced by diffusion. These samples had n-type conduction and their electrical resistivity varied in a wide range depending on the degree of compensation. For example, when the concentration of iron atoms was $1.5 \cdot 10^{16}$ cm^{-3}, the density of free electrons varied in the samples from 10^{16} to 10^{14} cm^{-3}, and according to Ref. 29 – even down to $3 \cdot 10^{13}$ cm^{-3}. Nevertheless, the ESR spectra of all the samples (recorded by different authors) were practically the same. A spectrum of this kind is shown in Fig. 21a (Ref. 13). This spectrum consists of one line with a wide base. The width of the base of this line varies with the orientation of the external magnetic field relative to the crystallographic axes of the

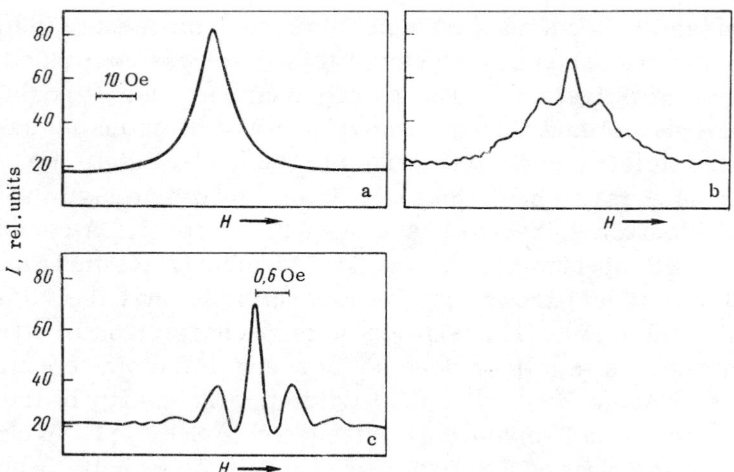

FIG. 21. Electron spin resonance spectrum of Si:Fe samples (H ‖ [111]; T = 1.25 K) obtained for microwave powers of 0.4 MW (a) and 40 MW (b, c).

sample. The base is widest when the magnetic field H is oriented parallel to the [111] direction and narrowest for H ‖ [100] (Ref. 27). Moreover, the line width is sensitive to the method used to prepare a sample. Careful investigations established[29] that this is due to internal mechanical stresses. In our opinion, a reliable confirmation of this conclusion is provided by the qualitative agreement between the angular dependences of the intensity and width of the ESR line of iron with the angular dependence of the Young modulus of silicon.[29]

It is clear from Figs. 21b and 21c that an increase in the microwave power splits the ESR line into several components separated by 0.5–0.6 Oe. This separation is independent of the external magnetic field.[30] Therefore, Feher[27] treated this splitting of the ESR line as a consequence of the hyperfine interaction of electrons with the ^{29}Si nuclei.

All the attempts to detect line splitting due to the crystal field of cubic symmetry have be unsuccessful and this indicates that the spin of the resonating iron atoms is $S \leq 3/2$.

The exact value of S was determined when the tetrahedral symmetry of silicon crystals doped with iron was distorted. This was achieved in Ref. 13 by mechanical compression of a sample and in Ref. 29 by grinding two faces of a sample, which was equivalent to uniaxial mechanical stretching at right-angles to these faces.

In both cases the spectral lines split into two components, indicating that $S = 1$. The correctness of this value was confirmed by an ENDOR investigation.[31] The results obtained made it possible to conclude firmly that the iron impurity in the investigated samples was located in tetrahedral interstices in the neutral state Fe^0. Had the Fe^0 configuration been the same as of the iron atoms in the free state, the electron spin would have been 2 and not 1. A comparison of the ESR spectra of Fe^0 and Mn^- demonstrates their qualitative similarity and shows that the resonating ion has the $3d^8$ configuration (Table IV). The position of zero-charge iron in tetrahedral interstices in silicon does not agree with the size considerations (Fig. 5) if we ignore the stabilization of the energy of the crystal lattice with T-metal ions, discussed in Sec. 1.1. We may assume that the shift of the Fermi level to the lower half of the band gap of an iron-doped silicon crystal makes the energy state of iron free of an electron, i.e. Fe^0 becomes Fe^+. It is indeed found that p-type samples containing not only iron but a shallow acceptor (boron, gallium, aluminium) exhibit the ESR spectrum of the Fe^+ ion.[32] The parameters of the spectra of Fe^0 and Fe^+ are given in Table VII. The Fe^+ state can be observed only when samples are cooled rapidly after short diffusion doping of crystals with iron. If a sample is cooled slowly, then the Fe^+ ions in silicon may form ion pairs with negatively charged shallow acceptors,[33] as discussed below.

Nickel

Nickel was the second T-metal in germanium investigated by the ESR method.[34] Samples were prepared by diffusion at 850° C. The concentration of nickel was $7 \cdot 10^{15}$ cm^{-3}. Before diffusion the samples contained a large number of shallow donors (arsenic) which filled the energy levels of nickel with electrons imparting a negative charge. Up to a certain arsenic concentration the intensity of absorption of microwave power was proportional to the concentration of nickel, but at higher concentrations the absorption fell by an amount which increased on increase in the arsenic concentration. This led to the suggestion[34] that the nickel ions in germanium are doubly charged. In the absence of arsenic the nickel impurity is mainly in the nonionized site state Ni^{4+}. In the presence of arsenic the nickel ions become compensated and the lower level of nickel is filled to form singly charged Ni^{3+} ions. At still higher arsenic concentrations the Fermi level approaches the bottom of the conduction band and the second acceptor level of nickel becomes filled so that all the nickel impurities assume the charge state Ni^{2+}. The corresponding reduction in the intensity of the ESR spectrum suggests that the spectrum reported in Ref. 34 was due to Ni^{3+} ions. The spectrum was observed with a good resolution at 20.4 K. At higher temperatures the spectral lines became broader and the resolution was lost.

The ESR spectrum of Ge:Ni^{3+} consists of several lines, the number of which may vary from two to six, depending on the orientations of the crystallographic axes of germanium relative to the external magnetic field. The ESR spectrum is characterized by an anisotropy of the g-factors. If the experiments are carried out so the [110] axis of a crystal is perpendicular to the magnetic field vector lying in a horizontal plane, four lines (A, B, C, and D) are observed and their amplitudes are in the ratio 1:1:2:2. Figure 22 shows the angular dependence of the g-factors for these lines in the spectrum.[34] The caption of this figure gives the principal values of the g-tensor: g_1, g_2, g_3. An analysis of this angular dependence shows that for any line in the spectrum the g-factor is given by the expression

$$g = (g_1^2\cos^2\theta_1 + g_2^2\cos^2\theta_2 + g_3^2\cos^2\theta_3)^{\frac{1}{2}}. \tag{16}$$

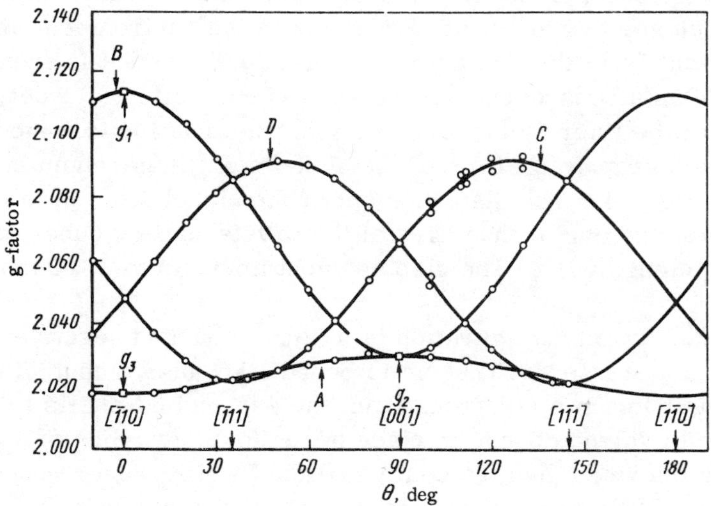

FIG. 22. Angular dependences of the g-factors of four lines (A, B, C, D) of the ESR spectrum of the Ni^{3+} ion in germanium (H⊥ [110]; T = 20.4 K); g_1 = 2.1128, g_2 = 2.0294, g_3 = 2.0176 (Ref. 34).

All the lines in the spectrum have the same profile. By way of example, Fig. 23 shows the B line from the spectrum. The similarity of the line profiles and the pairwise equality of their amplitudes led Ludwig and Woodbury[34] to the conclusion that each nickel atom contributes only one resonance transition ($S = \frac{1}{2}$), but there are six geometrically inequivalent sites at which the nickel atoms may be present. The distribution of these sites over a crystal is equiprobable.

The same authors[35] observed the hyperfine structure of the spectral lines for samples enriched with the ^{61}Ni isotope and the superhyperfine structure due to the interaction with the atoms of ^{73}Ge (nuclear spin 9/2)[34]. This investigation provided information very important for the understanding of the state of the Ni^{3+} ions in germanium. In fact, if we assume that nickel occupies the site positions in germanium (as indicated by the value of its spin found by the ESR method), then each atom of nickel is surrounded by four germanium atoms so that for any orientation of the magnetic field there should be four groups of superhyperfine structure lines. Each group corresponds to one principal line and should consist of $2I + 1 = 10$ lines of the superhyperfine structure. However, the experiments revealed only two groups of 10 lines. Hence, it follows that the nickel atom is displaced from a site so that it is closer to two out of four atoms in the environment and more distant from the other two. Such a displacement is treated in Ref. 34 as a consequence of the Jahn–Teller effect. A detailed analysis of the finer features of the spectrum has led to a model of the crystallographic position of the nickel atom in germanium shown in Fig. 24. The displacement of the nickel atom is in the direction of any one of the edges of the crystal lattice cube, for example, along [001]. The displacement amounts to 0.02 nm (Ref. 34).

The electron configuration of a singly charged negative nickel ion located at a site would seem to be $3d^7$, because a neutral nickel atom contributes two electrons from the 4s- and 3d-shells to the formation of valence bonds in place of the four electrons of a germanium atom which has the configuration $3d^6$. However, because of the Jahn–Teller effect, we can assume that a nickel atom forms bonds only with two germanium atoms and, therefore, the formation of these bonds requires only two electrons from the 4s-shell. In this case the configuration of nonionized and singly ionized nickel is $3d^8$ and $3d^9$, respectively. This leads to the spin of $\frac{1}{2}$, manifested experimentally by the ESR spectrum. A simple crystallographic

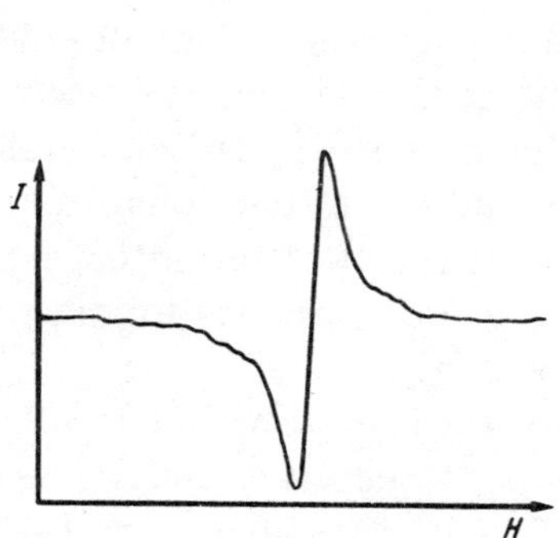

FIG. 23. Profile of the line B in the ESR spectrum of the Ni^{3+} ion in germanium (H ∥ [110]; T = 20.4 K) (Ref. 34).

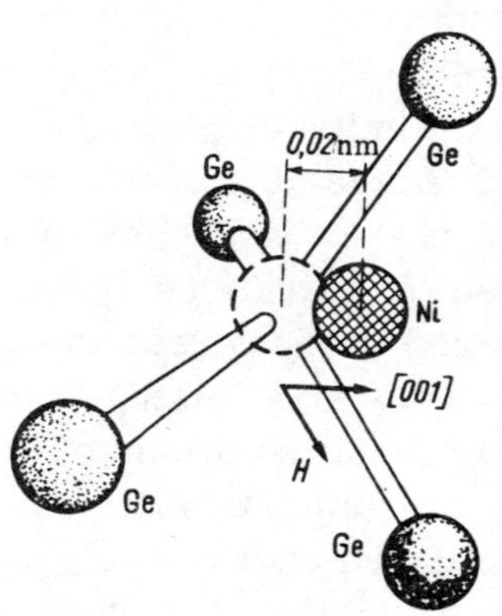

FIG. 24. Model of the state of a Ni^{3+} ion in the crystal lattice of germanium allowing for the Jahn–Teller shift.[34]

analysis (Fig. 6) shows that a nickel atom cannot possibly replace a germanium atom because of the difference in sizes. Clearly, Ge:Ni represents the case when a strong Jahn–Teller effect lifts the size forbiddenness by a change in the effective size of the T-metal ion.

The state of nickel atoms in silicon investigated by the ESR method was discussed in a review by Ludwig and Woodbury.[15] The parameters of the spectrum recorded at temperatures 10-20 K are listed in Table VII. Ludwig and Woodbury attributed the electron configuration $3d^9$ to the nickel ion on the assumption that it is located in a silicon interstice. However, it is said elsewhere in Ref. 15 that "... the Ni^{3+} ions occupy in silicon a position displaced relative to the lattice site, ...", i.e. there is an analogy with the behaviour of nickel in germanium. Moreover, cooling to 2 K or less lowers the rate of reorientation of the Ni^{3+} ions so that the ESR spectrum is split and, as concluded in Ref. 15, it is possible to distinguish several positions of the Ni^{+} ions which are geometrically inequivalent relative to the sites in the crystal lattice of silicon. The values of the g-factors corresponding to the crystallographic directions [111], [110], and [112] are 2.0133, 2.0357, and 2.0333.

It therefore follows that also in the case of silicon we can assume that the nickel atoms occupy the site positions but with a very large displacement because of a strong Jahn–Teller effect.

Unfortunately, the displacement in the case of Si:Ni has not been determined, but the large shift of the g-factor compared with Ge:Ni is an indirect confirmation of this hypothesis. The considerable displacement of the centres of the nuclei of the Ni^{3+} ions from the regular lattice sites in silicon allows us to regard such ions as located in interstices and assume that the wave functions are polarized in the direction of one of the crystal lattice sites especially as the size mismatch (Fig. 5) in no way permits formation of substitutional solutions of nickel in silicon.

The interstitial treatment of the position of nickel in silicon is supported also by the conservation of its positive charge as a result of cooling down to 1.05 K (Ref. 15). At these very low temperatures the usual impurities forming even shallow levels become deionized (this applies even more strongly to deep levels), i.e. nickel at a site should be neutral (Ni^{4+}). However, at an interstice a completely neutral state of nickel would have the configuration $3d^{10}$, but a strong coupling with one of the sites is ensured with one electron transferred to the valence band of silicon (irrespective of temperature) so that the nickel ion acquires the electron configuration $3d^9$ observed experimentally.

Model of States of Transition-Metal Impurity Atoms in Elemental Group IV Semiconductors.

The experimental results reported above show that T-metal impurities may be located at sites or in interstices of the crystal lattices of germanium and silicon. In both elements a T-metal atom is in a tetrahedral environment if the wave functions of the d-shell electrons are localized within the first coordination sphere. However, if the interaction of a T-metal atom with the second neighbours is strong, i.e. if the wave functions of the d-electrons extend to the second and possibly even further coordination spheres, the environment can no longer be regarded as simply tetrahedral. The problem of the degree of localization of the d-electrons of a T-metal atom in the lattice of silicon or germanium is considered in Ref. 36 using the crystal field theory. A model put forward as hypothetical by Woodbury and Ludwig[24] is used there: it is assumed that when a lattice site is occupied, the electrons from the 4s-shell and some of the missing (up to four) electrons from the 3d-shell of a T-metal atom are used to form bonds with the four nearest silicon atoms; when a T-metal atom is located at an interstice, electrons are transferred from its 4s-

TABLE VIII. Contributions of Different Coordination Spheres to Splitting $\Delta = E_e - E_{t_2}$ of T-Ion Levels in Silicon[36]

No. of CS	No. of atoms in CS	R_k/a	$(E_e - E_{t_2})/q\delta_\perp$, cm^{-1} for T substitutional atom	$(E_e - E_{t_2})/q\delta_\perp$, cm^{-1} for T interstitial atom
1	4	$\sqrt{3}/4$	–5880/–5880	–5880/–5880
2	12	$\sqrt{2}/2$	–1590/–1590	– 597/8050
3	12	$\sqrt{11}/4$	680/165	608/165
4	6	1	487/91.1	– 987/–248
5	12	$\sqrt{19}/4$	– 217/–34.2	– 266/34.2
6	24	$\sqrt{6}/9$	– 146/18.3	201/30.1

Note. 1. The numerator gives values corresponding to $q\delta_k$ = const and the denominator gives the values for the case when $q\delta_k \propto R_k^{-2}$. 2. Notation: CS denotes a coordination sphere; q is the charge of a T-metal ion; a is the lattice period; R_k and $q\delta_k$ are the radius and dipole moment of an atom in the k-th sphere, respectively, where $q\delta$ is measured in atomic units (a.u.).

shell to the 3d-shell.

An atom of the host lattice (silicon) is considered in Ref. 36 to consist of a positive point charge (nucleus) and an electron cloud which fills a sphere of radius ρ with a constant density. The centre of the sphere does not generally coincide with the position of the nucleus but is displaced by an amount δ in the direction towards a T-metal impurity atom, i.e. the electron density becomes polarized.

The reader interested in details of these calculations should consult Ref. 36; here, we shall consider the main results of this work which gives the relationship between the positions of the energy levels of a T-metal impurity in a crystal and the quantities ρ, δ, and α, where α is the familiar parameter of a one-electron radial function.[37] The main results of this theoretical calculation are presented in Table VIII for the least favourable case α = 1 a.u., which corresponds to the largest radius of the state.

It is clear from this table that the contribution to the splitting is governed by the first two coordination spheres. This result is important because it is frequently assumed (without any justification) that the d-electrons are strongly localized. Estimates of the contributions of the remaining four spheres to the splitting of a level indicate that they do not exceed the contribution of the first two spheres.

The main result of the theoretical analysis in Ref. 36 is the difference between the signs of the splitting Δ in the case of a T-metal impurity at substitutional and interstitial positions.

In the substitutional position (at a regular site) the value of Δ is negative, i.e. the energy of the e-states is less than the energy of the t_2-states. Hence it follows that the regions of lower electron density lie in the direction of the principal crystallographic axis of a cube. This is the hypothesis used by Ludwig and Woodbury[38] as the basis of their phenomenological model which is supported by the crystallochemical calculations in Ref. 36. Therefore, this model can be justifiably called the Roĭtsin–Firshteĭn–Ludwig–Woodbury model or, briefly, RFLW model. It also follows from Table VIII that when T-metal ions are at interstitial positions, the contribution of the second coordination sphere is opposite in sign to the contribution of the first sphere. The effect can be explained qualitatively by the fact that an increase in the energy of the e-states due to the influence of the second neighbours exceeds the relative reduction in the energy of these states as a result of the interaction with the nearest neighbours. Consequently, in this case the value of Δ is positive and the upper and lower energy states are t_2 and e, respectively.

This model makes it possible to explain very well the electron structure of the T-metal ions in silicon. This can be done if we assume that the electron occupancy of the orbitals of the e- and t_2-states obeys the Hund rule: the lowest energy is exhibited by the state which has the highest (for a given electron configuration) values of S and L.

In this case the predicted electron structure of the T-metal ions in silicon (Fig. 25) is supported completely by the experimental data given above. By way of a resumé of the model of states of the T-metal impurities in silicon, we can say that the general principle of the crystal chemistry of all the compounds of transition metals is confirmed in the interstitial states in group IV crystals: a transition metal tends to fill the d-shell in any possible way. The ions with the $3d^n4s^m$ configuration assume the $3d^{n+m}$ configuration in an interstitial state. This rule does not apply in a substitutional state. In the latter case the occupancy of the valence band of silicon or germanium predominates and ions with the $3d^n4s^m$ configuration assume the $3d^{n-(4-m)}$ configuration.

The RFLW model under consideration reflects qualitatively the physical situation of the state of a T-metal ion in the lattice of silicon or germanium. However, some effects cannot be explained even qualitatively by this model. This is particularly true of the Jahn–Teller effect manifested by a displacement of an ion from a substitutional position, producing a defect associated with a

$3d^n$	Occupancy of 3d orbit	Substitutional ions				Occupancy of 3d orbit	Interstitial ions			
$3d^1$	t_2 ≡ e =.	Sc^-	Ti^-	V^0	Cr^+	e = t_2 ≡.				Sc^{2+}
$3d^2$	≡ =:	Ti^{2-}	V^-	$\boxed{Cr^0}$	$\boxed{Mn^+}$	= ≡:			Sc^+	Ti^{2+}
$3d^3$	≡. =:	V^{2-}	Cr^-	Mn^0	Fe^+	= ≡⁝		Sc^0	Ti^+	$\boxed{V^{2+}}$
$3d^4$	≡: =:	Cr^{2-}	Mn^-	Fe^0	Co^+	=. ≡⁝	Sc^-	Ti^0	V^+	Cr^{2+}
$3d^5$	≡⁝ =:	$\boxed{Mn^{2-}}$	Fe^-	Co^0	Ni^+	=: ≡⁝	Ti^-	V^0	$\boxed{Cr^+}$	$\boxed{Mn^{2+}}$
$3d^6$	≡⁝ =:.	Fe^{2-}	Co^-	Ni^0		=: ≡⁝.	V^-	Cr^0	$\boxed{Mn^+}$	Fe^{2+}
$3d^7$	≡⁝ =::	Co^{2-}	Ni^-			=: ≡⁝:	Cr^-	$\boxed{Mn^0}$	$\boxed{Fe^+}$	Co^{2+}
$3d^8$	≡⁝. =::	Ni^{2-}				=: ≡⁝⁝	$\boxed{Mn^-}$	$\boxed{Fe^0}$	$\boxed{Co^+}$	Ni^{2+}
$3d^9$						=:. ≡⁝⁝	Fe^-	Co^0	$\boxed{Ni^+}$	
$3d^{10}$						=:: ≡⁝⁝	Co^-	Ni^0		

FIG. 25. Electron structure of T-metal ions in group IV semiconductors deduced using the RFLW model subject to the Hund rule (the "framed" states are those observed experimentally in silicon).

vacancy and, if the displacement is small, the association is not with the usual vacancy but with a fraction of a vacancy, as is found in the case of Ge:Ni (Fig. 24).

These defects have been ignored so far in the physics of semiconductors and it is not yet clear how to approach the description of such defects in terms of the energy spectrum of electrons in a crystal.

The RFLW model fails to forecast the quantitative characteristics of the states of T-metal ions in semiconductor crystals. In the final analysis this model is based on the more general crystal field theory. The importance of the work of Roĭtsin and Firshteĭn is that they show theoretically the validity of the model even in the case of some delocalization of the wave functions of the d-electrons,

TABLE IX. Parameters of ESR Spectra of Impurity Pairs Formed by T-Metal

Impurity pair	J	$g_\parallel$	$g_\perp$	D	F	a
				×10⁴		
$(Cr-З)^0$	5/2	2.0	2.0	806	62	–
$(Cr-{}^{27}Al)^0$	5/2	1.994	1.994	–740.1	60.64	–
$(Cu-{}^{69}Ga)^0$	5/2	1.995	1.999	86.8	73.5	–
$({}^{53}Cr-{}^{197}Au)^0$	3/2	2.015	1.999	$6.7 \cdot 10^4$	–	–
$({}^{55}Mn-{}^{11}B)^+$	5/2	2.004	2.004	510.6	21.0	17
$({}^{55}Mn-{}^{27}Al)^+$	5/2	2.003	2.006	185.1	20.1	20
$({}^{55}Mn-{}^{197}Au)^+$	3/2	2.003	2.006	$7.6 \cdot 10^4$	–	–
$({}^{55}Mn-{}^{67}Zn)^0$	5/2	2.001	1.996	685.0	59.6	30
$({}^{55}Mn-{}^{195}Pt)^0$	3/2	2.027	2.017	10^5	–	–
$({}^{55}Mn-{}^{197}Au)^-$	5/2	1.991	2.016	$1.4 \cdot 10^4$	–	–
$({}^{57}Fe-{}^{11}B)^0$	3/2	2.068	2.0	10^4	–	–
$({}^{57}Fe-{}^{71}Ga)^0$	1/2	5.087	2.53	–	–	–

Note. The indices 1 and 2 refer to a T-metal ion and to an acceptor, respectively.

but a rigorous quantitative determination of the splitting of levels Δ cannot be obtained on the basis of the theory of these authors because of the far too many simplifying assumptions. It must be pointed out that Roĭtsin and Firshteĭn stressed the approximate nature of their theory.[36] However, they did explain the sequence of the levels.

The Ludwig and Woodbury scheme (Fig. 25) depends on the smallness of the values of Δ compared with those typical of ionic crystals. Otherwise it would be difficult to fill the levels with electrons so as to obtain the maximum spin in accordance with the Pauli principle.

All this shows that the value of Δ should be found experimentally and the empirical data on the structure of the energy levels of the T-metal ions may be considered on the basis of the RFLW model of their state in the crystal lattice of silicon or germanium.

Associated States of T-Metal Ions

It is shown above that the ions of chromium, manganese, and iron

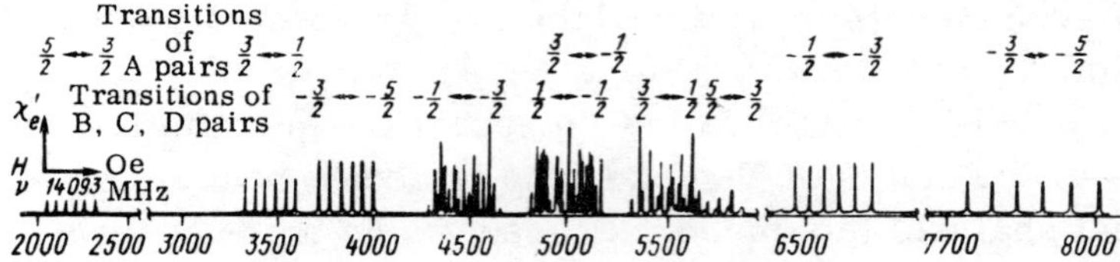

FIG. 26. Electron spin resonance spectrum of $(Mn-Zn)^0$ pairs in silicon.[15]

Ions in Silicon

A_1	A_2	B_1	B_2	p_1	p_2
–	–	–	–	–	–
–	–1.48	–	–	–	–
–	–4.93	–	–	–	–16.6
10.9	4.5	9.1	2 6	–	–
–52.5	–0.89	– 52.9	–1.8	–0.7	–0.05
–54.6	–1.81	– 53.1	– 1.27	–0.3	–0.37
–60.4	3.9	– 48.1	2.0	–0.9	–500
–51.4	– 1.0	– 50.0	–	–0.6	–
± 46.9	36.0	± 37.7	4.9	± 1.1	–
± 41.0	1.1	± 38.4	4.0	± 0.9	–
4.0	0.24	6.8	1.6	–	–
4.1	19.1	4.1	4.7	–	–

are located mainly in interstices of the silicon lattice. Such interstitial ions are characterized by a high mobility. Therefore, in the course of migration of these ions in a crystal they are likely to encounter ions of other impurities. The different charges of the ions which meet in this way give rise to the Coulomb interaction and formation of ion pairs. Such pairs were manifested in the ESR spectra of samples containing boron, gallium, aluminium, and indium acceptors, subjected to the subsequent diffusion of a T-metal impurity. The spectra of some pairs are shown in Figs. 26 and 27. It is clear from Fig. 26 that the spectrum of $(Mn-Zn)^0$ pairs oriented along the magnetic field (A pairs) consists of five groups of fine-structure lines ($J = 5/2$) and each of these is split into six lines by the hyperfine interaction with the nuclei of the ^{55}Mn isotope.

The spectra of pairs oriented with their [111] axis in a crystal

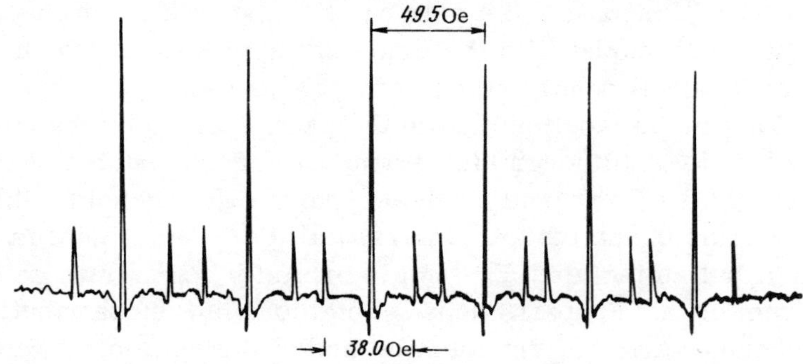

FIG. 27. Fine and hyperfine structures of the ESR spectrum of $(Mn-Pt)^0$ pairs in silicon.[15]

at an angle to the magnetic field (B, C, and D pairs) exhibit additional hyperfine structure lines due to the forbidden transitions. Figure 27 shows one of the three groups of the fine-structure lines observed in the ESR spectrum of $(Mn{-}Pt)^0$ pairs. The line is split into components by the hyperfine interaction with the nuclei of the ^{195}Pt isotope the abundance of which in a natural mixture of isotopes is 34%.

The symmetry of the distribution of the lines in the spectra has made it possible to establish reliably that the axes of all the investigated pairs, with the exception of $(Fe{-}In)^0$, are directed along the [111] axes in a silicon crystal, i.e. along the direction joining a lattice site to the nearest interstice. The parameters of the ESR spectra of all the pairs, with the exception of $(Fe{-}In)^0$, are therefore described well by the Hamiltonian (13). These parameters are listed in Table IX.

The axes of the $(Fe{-}In)^0$ pairs are parallel to the [001] axes joining a site not to the nearest but to the second nearest interstitial position. Ludwig and Woodbury[15] drew attention to the fact that for some pairs the parameters of the spectra (J, $g_{\parallel}$, $g_{\perp}$, a, A) differ only slightly from the parameters of the spectra of the corresponding T-metal ions in silicon that do not form pairs. This appears clearly if we compare the data in Tables VII and IX. These pairs include, for example, $(Mn{-}B)^0$, $(Mn{-}Al)^+$, $(Mn{-}Zn)^0$ with the spectral parameters very close to those of the spectrum of the interstitial Mn^{2+} ions. Similarly, the parameters of the spectra of the isolated interstitial Cr^+ ions are similar to the parameters of the $(Cr{-}B)^0$, $(Cr{-}Al)^0$, and $(Cr{-}Ga)^0$ pairs. Hence, it follows that the presence of an acceptor in a pair is manifested in practice only by the appearance of noncubic terms in the Hamiltonian. The formation of a pair can then be treated on the basis of a simple ionic model with the Coulomb interaction between a charged T-metal donor and any acceptor.

This model applies also to the pairs formed by the iron impurity in silicon although the parameters of the spectra of the $(Fe{-}B)^0$, $(Fe{-}Ga)^0$, and $(Fe{-}In)^0$ pairs differ considerably from the spectrum of the isolated interstitial Fe^+ ions. This is due to the fact that an isolated Fe^+ ion is orbitally degenerate so that the spectrum of the Fe pairs depends on the relative contributions of the noncubic terms of the potential and of the spin–orbit interaction. If the values of the potential are small, the total spin S and the effective orbital momentum L' of an ion are added so that in the

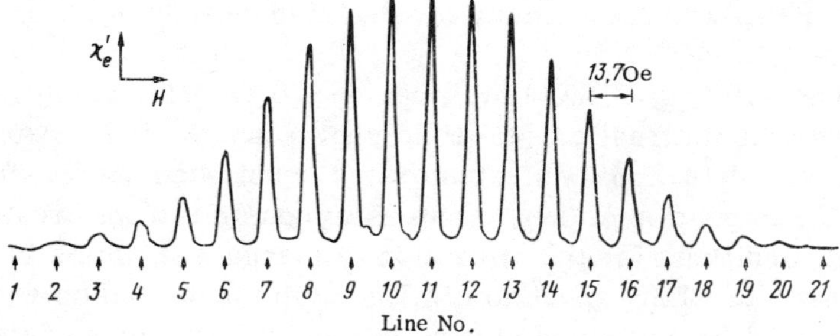

FIG. 28. Electron spin resonance spectrum of manganese clusters in silicon.[39]

ground state the Fe^+ ion has the momentum $J = \frac{1}{2}$. In this case the g-factor of a pair is anisotropic because of the noncubic terms in the Hamiltonian of Eq. (13). If the values of the potential are large, the orbital degeneracy of the Fe^+ ion is lifted and the parameters of the spectrum of impurity pairs correspond to $J = S = 3/2$ and $g = 2$ when the Hamiltonian contains noncubic terms.

The spectra of pairs formed by T-metal ions with other deep acceptors, for example, $(Mn-Au)^+$, are more complex[15] and the role of an acceptor does not reduce simply to introduction of noncubic terms. These pairs exhibit combination of the momentum $J = 5/2$ of the Mn^{2+} ion with the momentum $J = 1$ of the Au^- ion resulting in the pair momentum $J = 3/2$ (Ref. 15) and this demonstrates that the wave functions of the electrons of both partners become hybridized and the simple ionic model of formation of an impurity pair can no longer be applied.

Investigations of the ESR spectra of Si:Mn have revealed another associated state of the manganese impurity atoms, namely, a cluster. The ESR spectrum is then of the form shown in Fig. 28. The number of lines in the spectrum is 21, which corresponds to 21 possible values of m. Since $m = 2Jk + 1$, it is obvious that $k = 4$, i.e. the manganese impurity ions in the silicon lattice then form quartets. Unfortunately, without any additional study which has not yet been carried out, it is not known what is the size of such clusters so that it is not possible to say that a cluster of this kind represents a tetrahedral configuration of the manganese atoms that occupy the silicon sites in the first coordination sphere.

Another model of such a cluster is based on the interaction between manganese atoms located in different coordination spheres of the crystal lattice of silicon. This is known as the superexchange interaction and it is characterized by a very long range, for example, this interaction may involve manganese atoms in PbTe (Ref. 40).

The splitting of the ESR lines due to the Mn_4 clusters in silicon by the internal crystal field shows that $S = 2$. Attempts to observe ENDOR of Mn_4 clusters have not been successful.[13]

The formation of Mn_4 clusters is possible if we create favourable conditions for the migration of manganese atoms in silicon. Therefore, the ESR spectrum in Fig. 28 is observed only when manganese diffuses in samples with high or low values of the electrical resistivity but at a high temperature, followed by slow cooling.

1.4. STATE OF T-METAL IONS IN GALLIUM ARSENIDE

Iron

The reported ESR investigations of transition metal impurities in III-V compounds have been concentrated largely on the GaAs:Fe system.[41–46] This is clearly due to the fact that the paramagnetic absorption signal of this system is observed readily even at room temperature. This is characteristic of ions in the S state with a long relaxation time and it is due to the weakness of the interaction between the orbital singlet and the lattice. Moreover, the maximum solubility of iron in GaAs, reaching 10^{18} cm^{-3}, is much higher than the solubility of other transition metals in the same matrix. Finally, one should point out one other feature of the iron impurity which simplifies an analysis of the ESR spectrum. Of all the isotopes only the nucleus of ^{57}Fe has a nonzero magnetic moment (Table VI), but its natural abundance is only 2%. Therefore, there should be no hyperfine structure in the ESR spectrum of the Fe^{3+} ions, i.e. the third term in Eq. (8) should vanish.

The observed characteristic paramagnetic absorption spectrum consists of five quite easily resolvable lines (Fig. 29). The ratio of the line intensities in the spectrum is 3:5:10:5:3 (Ref. 45), which is practically identical with the calculated ratio. The angular dependence of the Fe^{3+} spectrum shown in Fig. 30 is described well by the spin Hamiltonian (8) with $A = 0$. The values of the parameters representing the ESR spectrum are listed in Table X.

The five fine-structure ESR lines give $S = 5/2$ and naturally the authors of all the cited papers attribute the observed ESR spectrum to the trivalent Fe^{3+} iron ions with the $3d^5$ configuration,

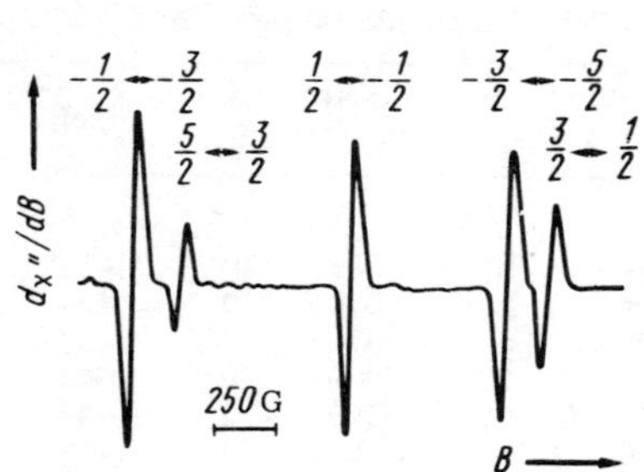

FIG. 29. Electron spin resonance spectrum of GaAs:Fe samples.

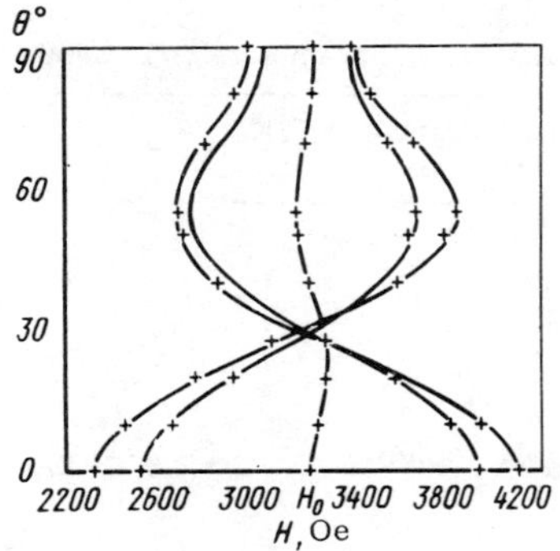

FIG. 30. Angular dependences of the fine structure lines in the ESR spectrum of iron ions in GaAs.

in agreement with Table IV. True, a similar spectrum is attributed to thermal defects in Ref. 43, but this conclusion is drawn only on the basis of a correlation between the concentration of these defects deduced from the ESR data and the concentration of the iron atoms found by mass-spectroscopic analysis. It is known that both the ESR and particularly mass-spectroscopic analysis very frequently fail to provide reliable quantitative data, in spite of the high sensitivity of these methods. For these reasons it is concluded in Refs. 45 and 47 that the conclusions reached in Ref. 43 on the nature of defects with spin 5/2 are incorrect. The $3d^5$ configuration leads necessarily to the conclusion that the Fe^{3+} ions form a substitutional solid solution and they occupy the sites in the Ga sublattice in the tetrahedral field of the matrix.

The preferential positioning of the iron impurity at the cation sites in gallium arsenide is supported also by a crystallochemical analysis given in Sec. 1 of the present chapter. Experimental investigations of GaAs:Fe crystals by the paramagnetic acoustic resonance (PAR) method were made on oriented single-crystal high-resistivity samples doped simultaneously with iron and a shallow donor impurity (tellurium) in concentrations of $5 \cdot 10^{17}$ and 10^{17} cm^{-3}, respectively.

It was found that these crystals exhibited a strong resonant absorption of longitudinal hypersonic waves of 9.4 GHz frequency (Fig. 31) at low temperatures (1.7-20 K). It was found that the absorption coefficient and the resonance magnetic field intensity depended strongly on the mutual orientations of the magnetic field, direction of propagation of sound, and crystallographic axes of

TABLE X. Parameters of T-Metal Ions in III-V Compounds Deduced from ESR Spectra

Ion	Configuration	T, K	S	J	g	ΔH_0, Э	$a \cdot 10^4$, cm^{-1}	A, cm^{-1}	Ref.
					GaAs				
Fe^{3+}	$3d^5$	1,3	5/2	5/2	2.046	54	339	10	[41]
		77			2.042	53	330	0	[42]
		77			2.05	55	370	0	[43]
		77			2.047	53	339	0	[44]
		300			2.05	53	356	0	[44]
Ni^{3+}	$3d^7$	1,3	3/2	9/2	2.106	130	0	0	[41]
		77			2.110	180	0	0	[45]
		100			2.114	150	0	0	[70]
Co^{2+}	$3d^7$	77	3/2	3/2	2.360	87	0	–	[45]
		100			2.356	90	0	0	[70]
Mn^{2+}	$3d^5$	77	5/2	5/2	2.004	30	3.75	60	[46]
		77			2.003	–	14	52.4	[72]
		77			2.0036	25–30	10	54.5	[73]
					Ga P				
Fe^{3+}	$3d^5$	10	5/2	5/2	2.025		390	–	[77]
		20			2.02			–	[78]
		100			2.025	40	390		[79]
Co^{2+}	$3d^7$	77	3/2	3/2	2.159	70	–	–	[80]
Mn^{2+}	$3d^5$	10	5/2	5/2	2.002	–	–	55	[77]
		77			2.003	–	10	53.8	[81]
		100			2.003	–	10	53	[74]
V^{2+}	$3d^3$	100	3/2	3/2	$g_\parallel = 2.02$ $g_\perp = 1.99$	140	–	–	[82]
					InP				
Fe^{3+}	$3d^5$	100	5/2	5/2	2.020	120	220	–	[83]
Mn^{2+}	$3d^5$	100	5/2	5/2	1.997	330	–	–	[88] [79]
					InAs				
Fe^{3+}	$3d^5$	4,2	5/2	5/2	2.035	130	420	–	[89]
		1,3			2.035	125	421	–	[90]

crystals.

An analysis of the results of a determination of the temperature and angular dependences of the principal characteristics of the resonant absorption of sound in GaAs:Fe showed that the observed relationships could be explained by transitions between the components of an excited spin–orbit triplet Γ_4 of the ^5E level of Fe^{2+} with the electron configuration $3d^6$ split by the applied magnetic field and an axial perturbation.[48]

The crystal field in gallium arsenide has the tetrahedral symmetry with a weak axial distortion along the [110] axis; this provides further evidence in support of the hypothesis that the iron impurity occupies the gallium lattice sites. The temperature dependences of the absorption and width of the resonance line were used to find the energy gaps between the ground and excited states

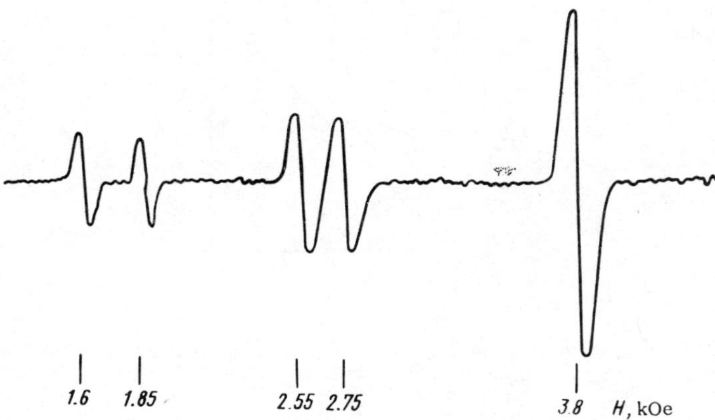

FIG. 31. Paramagnetic acoustic resonance of Fe^{3+} in GaAs (k_s ∥ [111]) (Ref. 48).

of the ^5E multiplet; these were 14.6, 33, and 50 cm^{-1}, respectively.

In addition to the strong hypersound absorption bands, the samples oriented along the [111] axis exhibited a structure consisting of five weak resonance lines of intensities two orders of magnitude less.[48] The profiles, the resonance fields, and the line width (~50 Oe) indicated that this PAR structure was fully analogous to the ESR spectrum of the Fe^{3+} ions in gallium arsenide.

The PAR data for the Fe^{2+} and Fe^{3+} ions made it possible to determine the electron–phonon coupling constant ε for the observed transitions and to find the absorption coefficient of hypersound in the direction of the twofold axis. The calculated values of ε were $1.1 \cdot 10^3$ cm^{-1} and 10 cm^{-1} for the impurity ions in the $3d^6$ and $3d^5$ configurations, respectively. The small value of ε for Fe^{3+} was due to the fact that the ground state ^6S was characterized by a very weak interaction with the lattice. This was supported also by the absence of a temperature dependence of the resonance line width of the Fe^{3+} ions in the case of ESR and PAR.

The PAR effect was used in Ref. 49 to study GaAs:Fe:Te samples and a simultaneous determination was made of the concentration of iron in the $3d^5$ state from the ESR spectrum and of the total concentration of both impurities by the method of ^{59}Fe and ^{123}Te radioisotope tracers and gamma-ray spectroscopy.[49]

In crystals specially selected for this purpose the concentration of iron varied within a narrow range of $(5\text{–}7) \cdot 10^{17}$ cm^{-3},

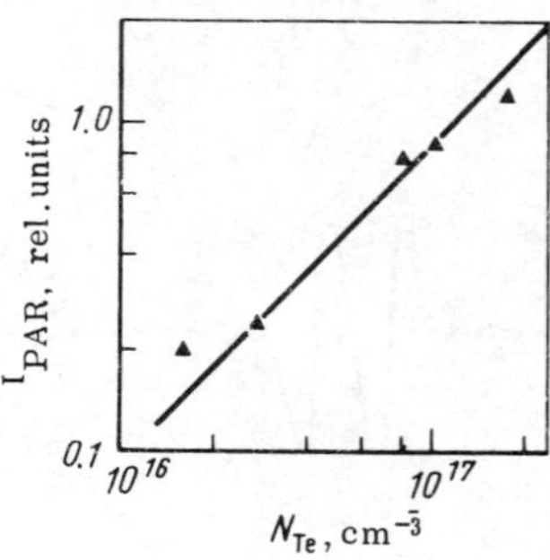

FIG. 32. Dependence of the intensity of the paramagnetic acoustic resonance signal on the concentration of the compensating donor (tellurium) in GaAs.[49]

whereas the donor impurity (tellurium) concentration varied within a wide range of $1.6 \cdot 10^{16}$–$1.7 \cdot 10^{17}$ cm^{-3}. The dependence of the absorption of hypersound by the $Fe^{2+}(3d^6)$ ions on the tellurium impurity concentration was determined (Fig. 32). The Fe^{2+} ion content was determined completely by the concentration of the compensating donors. The results obtained indicated that in GaAs crystals doped with iron up to concentrations of $(1\text{–}2) \cdot 10^{18}$ cm^{-3} practically all the iron occupied the gallium lattice sites and could assume the charge states Fe^{3+} (nonionized) and Fe^{2+} (singly ionized) with the electron configurations $3d^5$ and $3d^6$, respectively. The same charge states could be equally well denoted by Fe^0 and Fe^-, respectively.

The existence of iron in the Fe^{2+} state with the electron configuration $3d^6$ in GaAs:Fe:Te was confirmed by the optical absorption spectra[50,51] shown in Fig. 33 for a sample with the maximum Fe^{2+} ion concentration ($5.5 \cdot 10^{17}$ cm^{-3}).

This spectrum should be compared with the corresponding Tanabe–Sugano diagram calculated for $3d^6$ in a crystal field of tetrahedral symmetry (Fig. 11e). In this case the energy spectrum of the Fe^{2+} ions could be described by two parameters: the crystal field strength $\Delta = 10D_q$ and the spin–orbit coupling constant, which for a free ion in the ground state ^5D was $\lambda = -(100 \pm 10)$ cm^{-1}. According to the results of Ref. 26, the ^5E state corresponds to five equidistant levels separated by $6\lambda^2/\Delta$. Similarly the split-off multiplet $^5T_2$ consists of five levels. The fine structure of both multiplets is shown in Fig. 34.

In the limit $T \to 0$ only the lowest single level of the ^5E

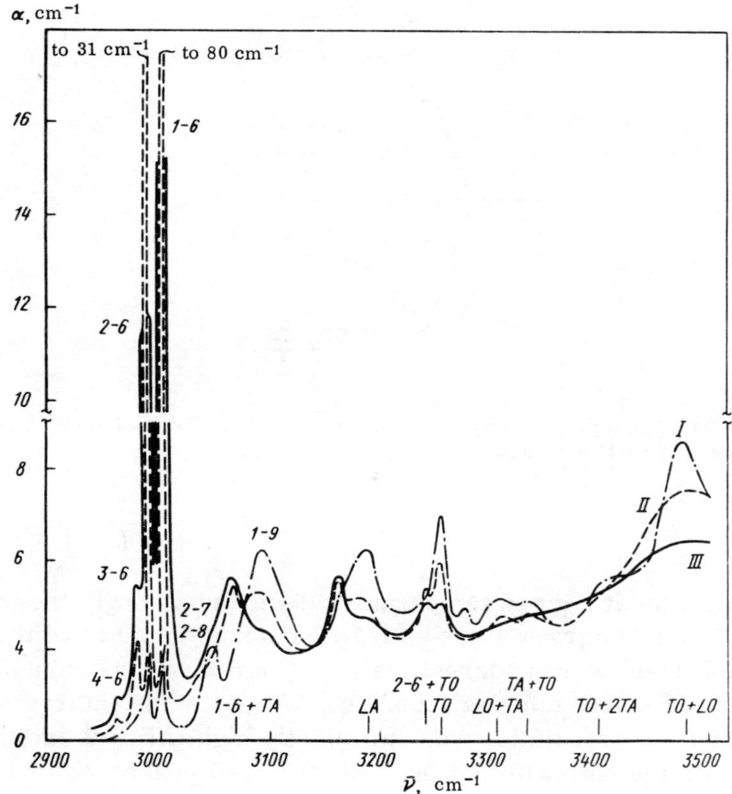

FIG. 33. Optical absorption spectra of GaAs:Fe at temperatures of 4.2 K (I), 10 K (II), and 20 K (III). The abscissa gives the values of the energies of the 1–6 transitions assisted by TA-, LA-, TO-, and LO-phonons and their combinations, as well as the phonon replicas of the 2–6 transition.

multiplet is populated; an increase in temperature causes a redistribution of electrons between the ^{5}E levels so that in addition to the 1–6 transitions, there should be also absorption peaks corresponding to the 2–6, 3–6, and other transitions. The intensities of these peaks are governed by the populations of the levels 1-5 and by the relevant selection rules. Clearly, this applies also to zero-phonon transitions 1–7, .., 5–7, 1–8, ..., 5–8, etc.

In fact, at T = 4.2 K (Fig. 33) only the absorption bands corresponding to 1–6 and 2–6, as well as 1–9 and 2–7 transitions are observed.

It was pointed out in Ref. 51 that insufficient resolution of

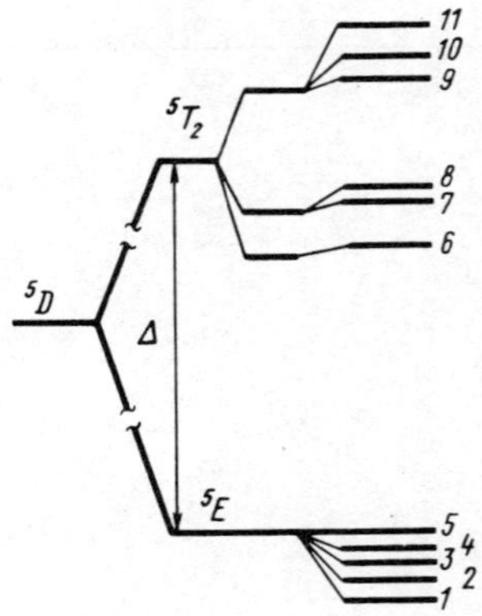

FIG. 34. Splitting of the atomic ^5D term in the tetrahedral field of ligands subject to the spin–orbit interaction.

the apparatus at a given temperature distorted strongly the profiles of the first two narrow lines so that the absolute values of the absorption coefficient were underestimated. Therefore, the values of α for T = 4.2 K are missing from Fig. 33. At temperatures T ≤ 10 K the resolution was sufficiently high ($\Delta\bar{\nu}$ = 2 cm^{-1}) compared with the half-width of the 1–6 and 2–6 peaks, which became much broader on increase in temperature. At temperatures 10 and 20 K there were also narrow absorption peaks corresponding to the 3–6 and 4–6 transitions. The 5–6 band, forbidden for symmetry reasons, was not observed in the spectra.

The positions of the maxima of all the identified zero-phonon peaks agreed for all the investigated samples and all temperatures to within 1 cm^{-1}. The experimental values are listed in Table XI.

The energy spectrum of the Fe^{2+} levels calculated in the

TABLE XI. Energies of Zero-Phonon $^5E-^5T_2$ Transitions in T-Metal Ions with $3d^6$ Configuration in GaAs

Transition $^5E \rightarrow {^5T_2}$	Energy, cm^{-1}	
	experiment[51]	theory[26]
1–6	3002	$\Delta + 3\lambda_T + 138\lambda^2/5\Delta$
1–9	3092	$\Delta - 2\lambda_T + 132\lambda^2/5\Delta$
2–6	2988	$\Delta + 3\lambda_T + 108\lambda^2/5\Delta$
2–7	3044	$\Delta + \lambda_T + 24\lambda^2/\Delta$
2–8	3044	$\Delta + \lambda_T + 30\lambda^2/\Delta$
3–6	2979	–
4–6	2962	–

crystal field theory approximation (Fig. 35) can be used in conjunction with the data in Table XI to determine the parameters Δ and λ.

One must allow for the influence of the dynamic Jahn–Teller effect, which – as shown in the case of the Fe^{2+} ions in II-VI compounds[52] – has little influence on the splitting of the ^5E term but may reduce significantly the spin–orbit interaction which is the main factor that governs the structure of the $^5T_2$ levels. This effect can be minimized by introducing[50] a correction coefficient denoted by λ_T in Table XI.

Subject to an allowance for the dynamic Jahn–Teller effect, the energies of the transitions, i.e. of zero-phonon peaks in the optical absorption spectra, can be linked to the splitting shown in Fig. 34 by the relationships given in the same Table XI.

A simultaneous solution of these relationships after substitution of the experimental values of the transition energies yielded[51] $\Delta = 2995 \pm 10\ cm^{-1}$, $\lambda = 77 \pm 6\ cm^{-1}$, and $\lambda_T = 18 \pm 2\ cm^{-1}$.

The results of optical spectroscopy thus confirmed the conclusion reached on the basis of ESR and PAR spectra that the iron atoms are located at the gallium sites in GaAs and have two charge states Fe^{3+} and Fe^{2+}, i.e. they exist in two electron configurations $3d^5$ and $3d^6$, respectively. The ratio of the concentrations of the iron ions in these states is controlled by external donors either present accidentally or specially introduced into a crystal. The conclusion on the occupancy of the gallium sites by

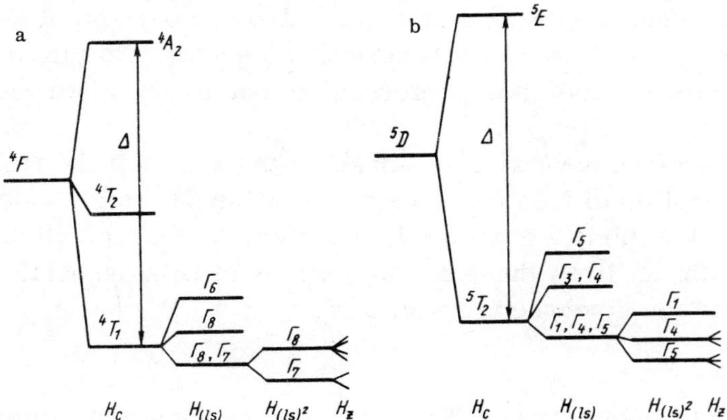

FIG. 35. Structure of the energy spectrum of the chromium atom in a crystal field of the tetrahedral symmetry T_d: a) $Cr^{3+}(3d^3)$; b) $Cr^{2+}(3d^4)$.

the iron atoms has also been drawn from the results of other investigations of GaAs:Fe by the Mössbauer effect[53] and by measuring the magnetic susceptibility.[54]

In view of this situation the results reported in Ref. 55 are surprising. Teuerle and Hausmann[55] investigated ENDOR in iron-doped GaAs and GaP single crystals and reached the unexpected conclusion that the Fe^{3+} ions are in the tetrahedral field of the matrix surrounded by the group V atoms but located in the tetrahedral interstices rather than at the Ga sites. In contrast to other methods, only the ENDOR technique can be used to distinguish (at least in principle) the location of an impurity centre in a tetrahedral interstice from that of an ion at a site of the T_d lattice (Sec. 1.2). Therefore, it would seem that the results of Ref. 55 cancel out the results of a large number of papers in which five different methods have been used. However, a detailed analysis of the data in Ref. 55 shows that they are quite insufficient for a reliable conclusion in support of one of the two possible models (impurities at the Ga sites or in the tetrahedral interstices). The nature of the angular dependence of the ENDOR frequency made it possible simply to establish the number of atoms and the symmetry of the coordination spheres surrounding an impurity atom. In both positions of the Fe^{3+} ions the nature of the nearest spheres is the same; the difference between them is only in the order and distances from the paramagnetic centre. Since the superhyperfine interaction constants have not yet been calculated allowing for the lattice deformation, a correct quantitative comparison of the theory with experiment (line intensity) cannot be made. This in turn means that we cannot give preference to one or the other model in this way.

All that we have said so far about the states of the iron atoms in GaAs applies to iron concentrations of the 10^{18} cm^{-3}, i.e. to the limit of solubility of iron. In the range of concentrations close to the solubility limit there are new states of iron associated with the interaction discussed in Sec. 1.7.

Chromium

All attempts to observe an ESR signal in various chromium-doped GaAs crystals at 300 and 77 K have been unsuccessful. Therefore, there are no data on the ESR in GaAs:Cr in the review of Bashenov.[45] The complex nature of the ESR spectrum of semi-insulating GaAs:Cr determined at 5 K and attributed to three different centres was already reported.[56,57] Part of the ESR spectrum with a strongly anisotropic (because of the Jahn–Teller effect) g-factor is explained by the presence of the Cr^{3+} ions at the Ga sites. It was found that the intensity of the spectrum decreased rapidly on increase in temperature, which was typical of systems with a strong spin–lattice coupling resulting in "smearing out" of the ESR lines at fairly low temperatures. It should be noted that for this reason a determination of the ESR spectra of the Cr^{2+} ions in CdS and ZnSe was made also below 4.2 K.

As pointed out in Sec. 2 of the present chapter, a strong electron–phonon interaction is a favourable factor for the observation of PAR. The resonance absorption spectra of hypersound of 9.4 GHz frequency[58] were compared with four possible models with the Cr^{3+} and Cr^{2+} ions forming substitutional or interstitial solid solutions with the crystal field symmetries T_d and O_h, respectively. Figure 35 shows the energy level scheme for these ions in the case of the T_d symmetry of the surrounding atoms in a crystal.[58]

The ground states $^4F(Cr^{3+})$ and $^5D(Cr^{2+})$ are degenerate and in a crystal field with T_d symmetry they split in accordance with the schemes shown in Fig. 35. This figure includes also a scheme for lifting degeneracy as a result of the spin–orbit interaction in the first and second orders of perturbation theory, the Zeeman splitting in a magnetic field, and the action of fields of symmetry lower than T_d which appear because of the instability of the orbital triplet to small deformations of the ligands.

An analysis of the energy structure of chromium impurity ions given in Fig. 35 makes it possible to estimate the feasibility of experimental investigation of these ions by resonance methods. The ground state of the chromium ions in both configurations is an orbital triplet. It follows from the Jahn–Teller theorem that this triplet splits, providing an opportunity for phonon-assisted transitions with a given spin between the Jahn–Teller sublevels. The occurrence of these transitions may result in strong broadening of the ESR lines and make it difficult to observe ESR in GaAs:Cr, as found experimentally.

The experimentally determined dependences of the absorption of sound on the temperature of a crystal, and on the intensity and direction (relative to the crystallographic axes) of the applied magnetic field[58] can be explained only by induced transitions between the split levels of a spin–orbit triplet of the term $^5T_2$ of the Cr^{2+} ion located in a field of the T_d symmetry. It is natural to account for these results by postulating that the chromium ions occupy the Ga sites in gallium arsenide, by analogy with the behaviour of iron in these crystals. One electron is then transferred from the 4s-shell and two electrons from the 3d-shell to form valence bonds with the anion (arsenic) electrons. In this case the nonionized Cr^0 chromium atom or, equivalently a Cr^{3+} atom, has the electron configuration $3d^3$. Since a crystal contains accidental donor impurities, part of the introduced chromium is compensated and assumes the ionized state Cr^{2+} of the $3d^4$ configuration.

However, the size mismatch discussed in Sec. 1 of the present chapter forbids the replacement of gallium with the chromium ions. This conflict is removed if we assume that a strong Jahn–Teller effect acts in GaAs:Cr. In fact, as established in a study of PAR (Ref. 58), the fine structure of the energy spectrum (Fig. 35) is affected significantly by the Jahn–Teller effect.

The hypothesis that chromium occupies the Ga sites in gallium arsenide was checked by optical spectroscopy.[59] Experiments were carried out on single crystals grown from the melt. Two groups of samples were investigated. Those belonging to the first group were doped only with chromium in concentrations of $(0.8-3)\cdot 10^{17}$ cm^{-3}. The concentration of the background of the residual shallow donors N_d^* was $(5-15)\cdot 10^{15}$ cm^{-3}, i.e. practically the whole of the dopant was in the neutral state with the electron configuration $3d^3$. Crystals of the second group contained not only chromium but also shallow donor centres of tin in order to create a controlled compensation of deep chromium levels which determine the ratio of the chromium impurity concentrations in the Cr^{3+} and Cr^{2+} states. The concentrations of the chromium and tin impurities in these samples were determined by the radioactive tracer method.

Figure 36 shows the optical absorption spectra of crystals of both groups recorded at 300 K. The results obtained show that the absorption in samples of the first group containing chromium mainly in the Cr^{3+} state was considerably greater than the absorption in samples of the second group where the dopant was in the Cr^{2+} state. This could be explained by the selection rules governing optical transitions of electrons in the d-shell of the transition metal impurity to the s-like bottom of the conduction band and to the p-like top of the valence band. The former d–p transitions are forbidden by the parity rules, whereas the latter are allowed.

This division into the allowed and forbidden transitions is fairly arbitrary because there are certain factors which at least

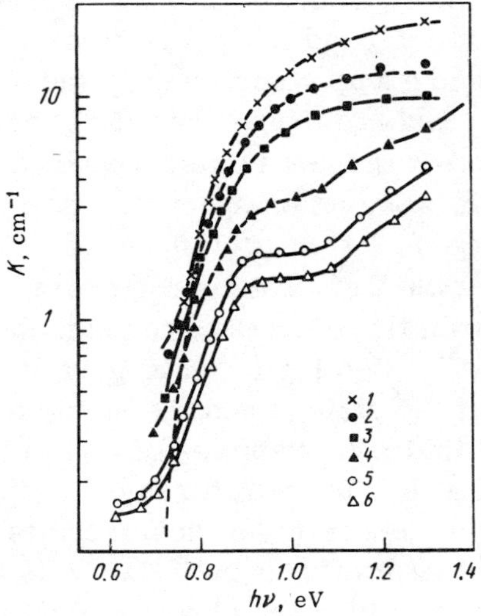

FIG. 36. Spectral dependences of the absorption coefficient of GaAs:Cr at 300 K obtained for samples with different concentrations of chromium N_{Cr} and of a donor N_d (both 10^{17} cm^{-3}): 1) 3.3 and 0.05; 2) 2.7 and 0.08; 3) 1.4 and 0.07; 4) 1.5 and 1.3; 5) 1.36 and 1.55; 6) 1.07 and 1.20. The dashed curve represents a theoretical dependence of the ionization cross-section on the photon energy.[60]

partly can lift the forbiddenness of the exchange of the d-states with the conduction band. Nevertheless, the cross-sections of the photoionization of an electron q_n and a hole q_p from a chromium level are such that $q_n/q_p \ll 1$.

Measurements in a large number of samples demonstrated that those belonging to the first group exhibited an increase in the impurity absorption in direct proportion to the Cr^{3+} content. In the case of the samples of the second group the dependence of the absorption coefficient on the Cr^{2+} concentration was linear only when the Cr^{2+} concentration was high; at low concentrations of Cr^{2+} the absorption varied slightly indicating a considerable contribution to the measured net absorption of some outside mechanism not associated with chromium, but governing partly the absorption in samples of the second group when the concentration of the chromium impurity in the Cr^{2+} state was low. The influence of optical transitions between a chromium level and a valence band was avoided in these measurements by employing mainly overcompensated (low-resistivity) samples of GaAs:Cr:Sn characterized by $N_{Sn} > N_{Cr}$ and by the equilibrium Fermi level located near the bottom of the conduction band. The additional absorption in the samples of the second group could be explained by optical transitions between deep levels of oxygen and the conduction band. It is clear from Fig. 36 that there was also a qualitative difference between the form of the absorption spectra of samples from the two groups. In the range $0.9 < h\nu < 1.05$ eV the dependence $K(h\nu)$ for GaAs crystals with Cr^{2+} exhibited a characteristic plateau, which was absent from the spectra of the samples from the first group containing Cr^{3+}. Cooling to 77 K gave rise to an absorption peak in the region of the plateau and the maximum of this peak was located at $h\nu_{max} = 0.91$ eV. Samples of the first group did not exhibit this peak even at low temperatures.

It is possible to deduce from these results that the absorption impurity peak exhibited by GaAs at 0.91 eV is not related at all to the chromium impurity but to the chromium atoms in the Cr^{2+} state. The resonant nature of this peak at low temperatures is due to the intracentre optical transitions between the $^5T_2$ and ^5E states into which the ground term ^5D of the Cr^{2+} ion splits in the tetrahedral crystal field of gallium arsenide (Fig. 35). The peak also explains the observation that the absorption spectrum of the samples of the first group could be described satisfactorily by the theory of Lucovsky[60] for the impurity absorption (dashed curve in Fig. 36). A similar curve calculated for the samples of the second group did not provide agreement which could in any sense be regarded as satisfactory.

It was interesting that in addition to one wide resonance peak the spectrum showed no fine structure which might be expected on the basis of a similar action of the crystal field on the $d^4(Cr^{2+})$ and $d^6(Fe^{2+})$ configurations, as demonstrated by comparison of the corresponding Tanabe–Sugano diagrams (compare Figs. 11c and 11e).

It should be noted in this connection that chromium-doped II-VI crystals exhibit a fine structure in the absorption spectrum and this structure is interpreted by electron transitions between the $^5T_2$ and ^5E multiplets.[61]

The absence of a fine structure in the absorption spectrum of GaAs:Cr led to the conclusion[59] that this material exhibits a rare but very interesting situation. In contrast to wide-gap II-VI crystals, in which both the $^5T_2$ and ^5E terms are in the band gap, the excited ^5E state of a defect in GaAs lies within the continuous spectrum of the conduction band, i.e. it is in resonance with the continuum. The resultant interaction between the discrete state and the states in the continuum should not only result in a significant delocalization (broadening) of the impurity level, but it should also perturb the states in the continuous spectrum near the energy of this pseudo-local level. This broadening destroys the expected fine structure of the spectrum because the degree of delocalization of the excited ^5E state is greater than the splitting of both terms by the spin–orbit interaction and other weaker effects.

Fano was the first to consider theoretically the role of a resonance between an excited discrete level and continuous states in the absorption spectra in the specific case of the absorption by ions of the rare gases.[62,63] It was found that an allowance for the interaction between various electron states of the ions results in an instability and autoionization of the discrete states. In isolated systems of this type the interaction of electrons inside one isolated centre gives rise to configurational mixing. We shall consider later an interesting feature of this situation: an excited state of the chromium impurity centres in a GaAs crystal is unstable against the formation (autoionization) of a free electron only because of the interaction of electrons in the impurity centre with the electrons of the host atoms. Later the ideas on the interchannel interaction and autoionizing states have been applied to solid-state physics in theoretical descriptions of the exciton absorption[64] and of optical transitions of impurities to a state in resonance with the continuum.[65]

Figure 37 shows the absorption spectra obtained at various temperatures for a low-resistivity GaAs:Cr:Sn sample which were identical with the spectra of semi-insulating crystals of the second group, except that the high-resistivity samples did not exhibit a weak absorption of light by free electrons in the photon energy range $h\nu < 0.7$ eV. This low-resistivity sample showed no contribution of the stronger transitions from the valence band to the chromium level. The results in Fig. 37 demonstrate that cooling hardly affected the energy position of the impurity band maximum, but it altered its half-width and this was clearly associated with the electron–phonon interaction. The spectrum as a whole was a result of a superposition of two bands or two absorption channels corresponding to optical transitions of electrons from the $^5T_2$

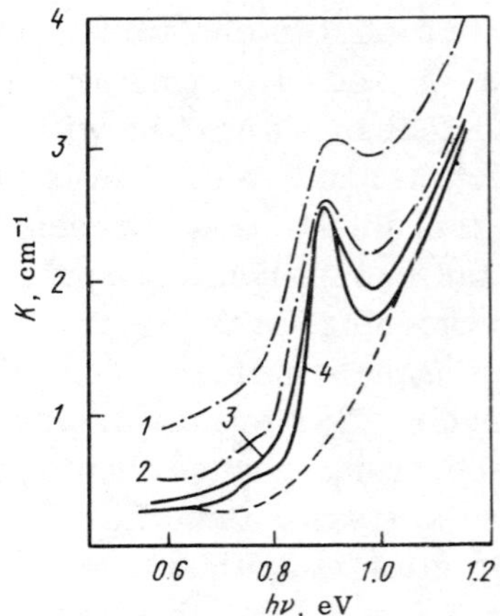

FIG. 37. Spectral dependences of the absorption in sample No. 6 (see Fig. 36) at the following temperatures T (K): 1) 300; 2) 77; 3) 20; 4) 4.2. The dashed curve represents an approximation of the optical absorption from the chromium level to the continuum.

state to the pseudo-local excited state ^5E which was in resonance with the continuum and to the continuous spectrum, as well as of the absorption due to electron transitions from deep levels of a residual defect to the conduction band.

There is experimental evidence of the existence of pseudo-local states of defects in some II–VI crystals doped with titanium, vanadium, and cobalt impurities.[66,67] However, the idea of a resonance of local states with the continuum[62] has not been used in these investigations nor in the study of the absorption spectra in the GaAs:Cr system.[68,69] This is a pity because the absorption band profile can be used to determine some important parameters of the investigated system even if the actual form of the wave functions of a defect is not known.

These parameters include particularly the fundamental characteristic of an impurity centre which is the matrix element of the interaction of a pseudo-local state with the states in the conduction band, which governs the degree of the hybridization of these states. It would be highly interesting to study the influence of a resonance on the probability of intracentre transitions. It is shown in Ref. 59 that hybridization of an excited impurity state with the states in the continuum results in some relaxation of the parity forbiddenness of the d–d transitions.

According to Fano's theory, the interaction between a discrete state φ in resonance with free $\psi(E)$ states in the continuum results in delocalization of the discrete state which then forms an energy band; the density of states in this band is described by a Lorentzian curve with a half-width $\pi|V(E)|^2$, where $V(E)$ is the matrix element of the hybridization of φ and $\psi(E)$. It is important to note that this interaction results in autoionization of the discrete state. If at some moment in time the system is in the state φ, it follows from the principle of indeterminacy that it becomes autoionized after the average lifetime of a resonance state $\tau \sim \hbar/\pi|V(E)|^2$. In the case of GaAs:Cr this means that the electrons released by the absorbed light from the ground state $^5T_2$ to the excited state ^5E become free after a time τ and this is followed by a rapid thermalization at the bottom of the conduction band. These are the electrons that are responsible for the impurity photoconductivity peak exhibited by GaAs:Cr.

Estimates of the lifetime of a resonance state were obtained in Ref. 59 and it was found that this lifetime is $\sim 10^{-14}$ sec. Moreover, the value of the matrix element of the hybridization of a discrete excited state ^5E of the Cr^{2+} ion with states in the conduction band of GaAs, $|V_E| \sim 0.1\ eV^{\frac{1}{2}}$, was deduced.

Hence, it follows that the hybridization energy of $|V_E|^2 \sim$ ~ 0.01 eV is much greater than the splitting λ of the ^5E and $^5T_2$ terms by the spin–orbit interaction. This clearly accounts for the absence of the fine structure from the optical absorption spectrum of GaAs:Cr.

Nickel

The state of the nickel impurity in GaAs was investigated by the ESR method[45,70] using single crystals to which nickel was added during growth. Chemical analysis data indicated that the concentration of nickel was $3\cdot10^{16}$–$1\cdot10^{17}$ cm^{-3}. The samples had p-type conduction and a fairly high resistivity (10–$10^4\ \Omega\cdot$cm). The ESR spectrum recorded at 100 K consisted of a single line. Its parameters are listed in Table X. The same table includes earlier data,[41] which are not very different. The simple line profile is typical of ions which have an orbital singlet as the ground state. The intensity of the spectral line decreased on increase in the degree of compensation of the samples and in the case of the more strongly compensated samples with a resistivity of $10^4\ \Omega\cdot$cm it was not possible to record the ESR spectrum at all. This circumstance together with the conclusions deduced from the crystallochemical analysis led to the hypothesis that nickel in GaAs occupies the Ga sites and that it has the $3d^7$ configuration in the neutral state. An increase in the degree of compensation by donors results in the conversion of an increasing proportion of the Ni^0 (Ni^{3+}) atoms to the Ni^- (Ni^{2+}) charge state with the $3d^8$ configuration. The $3d^7$ state corresponds to the quantum numbers $L = 3$, $S = 3/2$, and $J = 9/2$. The ground state of this configuration splits in a tetrahedral crystal field of gallium arsenide into a series of levels (Fig. 11f). We may assume that at the low temperature at which the ESR was investigated (100 K) only the lowest of the $^4A_2$ levels is populated. This level is characterized by $L = 0$ and in an external magnetic field it splits into four sublevels with different projections of the spin moment: $\pm 3/2$, $\pm 1/2$. The experimentally observed ESR spectrum is the result of electron transitions between these sublevels and it is described by just the first term of the Hamiltonian (8) with the effective spin $S = 3/2$. The second term in Eq. (8) disappears because of the absence of splitting in a crystal field of cubic symmetry. The ^{58}Ni and ^{60}Ni isotopes with a combined natural abundance of almost 100% do not have a magnetic moment so that the third term in Eq. (8), representing the hyperfine interaction with the T-metal nuclei, vanishes in the case of the Ni^{3+} ions.

The optical absorption spectra of GaAs:Ni are very different in the case of the Ni^{3+} and Ni^{2+} states. These spectra are shown in Fig. 38 (Ref. 71). The degree of compensation of GaAs:Ni samples was deduced from a comparison of the results of a spectrochemical determination of the total concentration of nickel with the ESR spectra which could be used to find the concentration of the Ni^{3+} ions. It is clear from Fig. 38 that in the case of GaAs:Ni^{3+} (curve 1) there is a wide absorption band typical of the photoionization of deep impurity centres in semiconductors. It is pointed out in Ref. 71 that the intensity of this absorption band varies with the nickel concentration and is correlated with the intensity of the ESR line.

Single crystals of GaAs:Ni strongly compensated by the residual donors (curves 2 and 3) or as a result of double doping with both nickel and tellurium (curve 3) exhibit another absorption band with a maximum at $\hbar\omega = 1.15$ eV. It is characteristic that after high-temperature treatment of samples 2 and 3 (annealing for 4 h at 1100° C followed by quenching in ethylene glycol) the result was decompensation of the samples, which increased greatly the intensity of the ESR line and of the optical absorption band

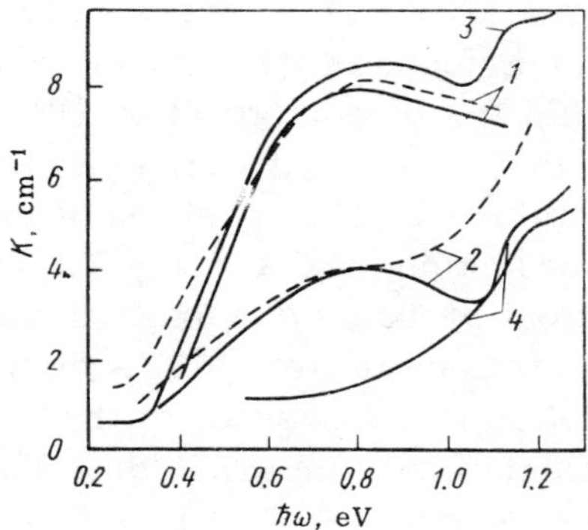

FIG. 38. Absorption spectra of GaAs:Ni samples recorded at 77 K (continuous curves) and at 300 K (dashed curves). The numbers alongside the curves represent samples with different nickel concentrations in the Ni, Ni^{3+}, and Ni^{2+} states, respectively (10^{16} cm^{-3}): 1) 5.5, 5.0, 0.5; 2) 10, 2.9, 7.1; 3) 16, 5, 11; 4) 13, 0, 13.

associated with the photoionization of a deep level, but in the region of 1.15 eV there were no singularities in the absorption spectrum. These observations and the small width (0.2 eV) of the band at 1.15 eV led the authors of Ref. 71 to draw the conclusion of intracentre electron transitions within the d-shell of the d^8 configuration of the compensated Ni^{2+} ions responsible for the absorption band at 1.15 eV.

The experimentally observed optical absorption spectrum is in good agreement with a model of the $Ni^{2+}(3d^8)$ ion in a crystal field of the tetrahedral symmetry. In fact, it follows from the appropriate Tanabe–Sugano diagram (Fig. 11g) that the ground state ^3F of this ion splits into three multiplets $^3T_1$, $^3T_2$, and $^3A_2$. The spin-allowed transitions are those from the ground state $^3T_1(F)$ to the excited states $^3T_2(F)$, $^3T_1(P)$, and $^3A_2(F)$. It follows from the crystal field theory that the energy gaps between the ground $^3T_1$ and the excited states of the d^8 configuration are given by the expressions in Table V.

The task is to identify the 1.15 eV absorption band with one of these transitions. This can be done by estimating the values of Δ and B. The former can be found from the dependence of Δ on the degree of ionicity (on the Phillips scale) of the crystal matrix given by Eq. (5) (Ref. 8). In the case of the Ni^{2+} ion the value of

Δ is 5600 cm^{-1}. The Racah parameter B can then be determined from the dependence given by Eq. (5) by introducing the refractive index of GaAs (n = 3.4), which yields B = 310 cm^{-1}. Substituting the values of Δ and B found in this way into the expressions for the transitions within the d^8 configuration (Table V), we find that the energy gap for the $^3T_1 \rightarrow {}^3T_1(P)$ transition is 1.15 eV.

It therefore follows that the position of the experimentally observed optical absorption band at $h\nu$ = 1.15 eV is such that we can identify it as most probably due to a transition of the $^3T_1(F) \rightarrow$ $\rightarrow {}^3T_1(P)$ type, as was done in Ref. 71. A distinguishing feature of this electron transition is the position of the excited $^3T_1(P)$ state in the conduction band of gallium arsenide. This is analogous to the behaviour of the ^5E state of Cr^{2+} in gallium arsenide. Since the $^3T_1(P)$ state of Ni^{2+} resonates with levels in the continuous spectrum, the transition in question should result in charge transfer (of electrons from the conduction band) and should be manifested in the steady-state photoconductivity spectrum. In fact, the experimental results of Ref. 71, which are plotted in Fig. 39, reveal a dip in the photoconductivity spectrum at $\hbar\omega$ = 1.15 eV in the case of the compensated samples in which the nickel ions are mainly in the $Ni^{2+}(3d^8)$ state. Conversely, the uncompensated samples in which practically all the nickel ions are in the $Ni^{3+}(3d^7)$ state there is no such dip (curve 1 in Fig. 39).

The multiplets formed by the crystal-field splitting of the

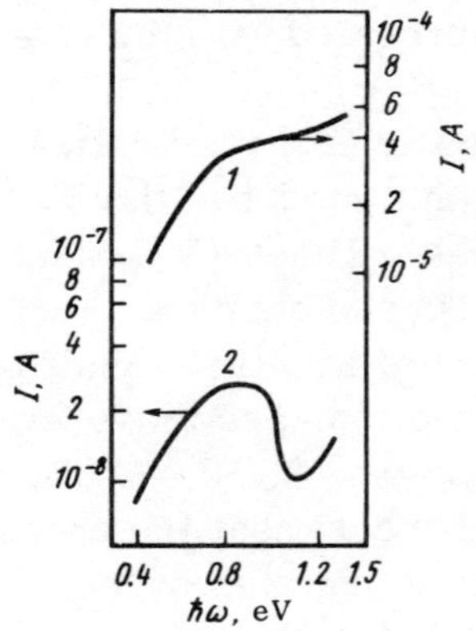

FIG. 39. Photoconductivity spectra of GaAs:Ni samples recorded at 80 K (1) and 200 K (2). The numbers of the curves are the same as the numbers of samples in Fig. 38.

ground state ^3F of the Ni^{2+} ion are themselves subject to further splitting by the spin–orbit interaction and this should, in principle, result in a fine structure of the optical absorption curve. However, the experimental results indicate (Fig. 38) that there is no such fine structure. Clearly, as in the case of the $Cr^{2+}(3d^4)$ ion, the interaction of the d-shell of the nickel ion in the $^3T_1(P)$ state with the states in the continuum of the conduction band of gallium arsenide is stronger than the spin–orbit interaction. The fine structure of the spectrum is therefore smeared out even at helium temperatures.

Cobalt

In contrast to nickel, the ESR spectrum of cobalt in GaAs is exhibited only by strongly compensated samples.[45,70] If we assume that cobalt also forms a substitutional solid solution in GaAs and occupies the sites in the Ga sublattice, then its configuration should be $3d^6$. However, in the case of strong compensation the cobalt ions acquire a charge and their configuration becomes $3d^7$. Such Co^{2+} ions are isoelectronic analogues of Ni^{3+}. Therefore, the ESR spectrum consists of a single line in both cases. The parameters of such a line are given in Table X.

In general, the ESR spectrum of the Co^{2+} ion should exhibit, in accordance with the Hamiltonian (8), the hyperfine interaction because the spin of the ^{59}Co nuclei characterized by a natural abundance of 100% is 7/2 (Table VI). Consequently, there should be seven hyperfine structure lines. However, it is concluded in Ref. 70 that the hyperfine structure is not resolved because of a strong interaction with the ligands.

By analogy with GaAs:Ni:Te, Andrianov et al.[70] prepared samples which were double doped with Co and Te. The cobalt impurity was entirely in the $Co^{2+}(3d^7)$ state. The optical absorption spectra of these samples exhibited a structure characteristic of electron transitions from the $^4A_2(F)$ ground state to the $^4T_1(P)$ excited state in accordance with the Tanabe–Sugano diagram (Fig. 11f) and with the data in Table V.

Manganese

Investigations of the ESR of manganese ions in GaAs have been made on samples with sufficiently large numbers of donors. The ESR spectrum (Fig. 40) consists of six lines which can be regarded as manifestations of the hyperfine structure.[46,72,73] This structure should be exhibited by the manganese ions because the natural

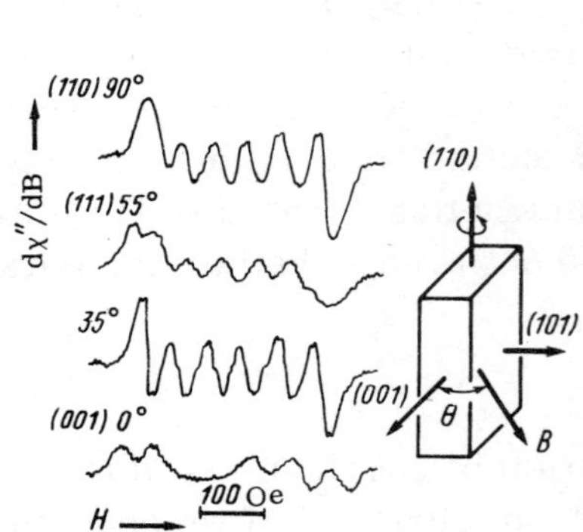

FIG. 40. Hyperfine structure of the ESR spectrum of GaAs:Mn at T = 77 K (Ref. 46).

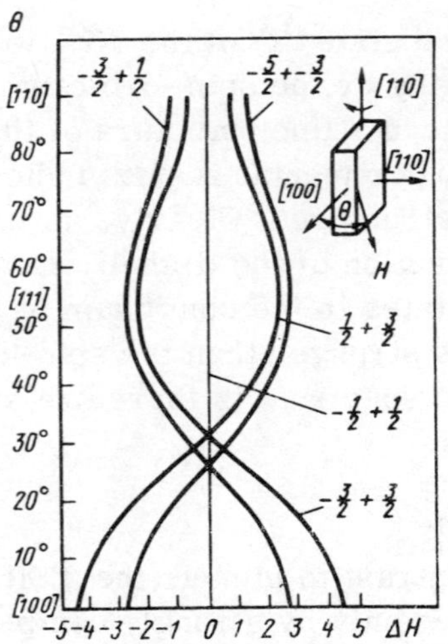

FIG. 41. Angular dependences of the fine structure lines of the ESR spectrum of GaAs:Mn (Ref. 46).

abundance of the ^{55}Mn isotope with the nuclear spin I = 5/2 is 100% (Table VI). The fine structure lines are found to be anisotropic and their width is independent of temperature in the interval 4-77 K. The anisotropy of these lines is explained in Ref. 46 by assuming that the hyperfine and the fine-structure lines are superimposed. Calculations of the frequencies of the transitions corresponding to the fine structure have been made as a function of the angle θ between the [100] axis of a crystal and the magnetic field (Fig. 41). It is clear from this dependence that the hyperfine structure is best resolved for those values of θ characterized by the weakest fine structure lines. Figure 41 demonstrates a good agreement with this conclusion. The fine structure of the ESR spectrum of GaAs:Mn consists of five lines, which corresponds to S = 5/2 and, as can be seen from Table IV, the electron configuration is $3d^5$, i.e. the Mn^{2+} ion is in a tetrahedral crystal field. Hence, the manganese ions occupy the gallium sites giving rise to the $3d^4$ configuration. This is followed by the capture of an electron from the donors present in a crystal and the transfer of the manganese atoms to the $3d^5$ configuration. We recall that the ESR spectrum of GaAs:Mn is observed only in the presence of a deliberately introduced donor.

A crystallochemical analysis (Sec. 1 in the present chapter) shows that the manganese ion cannot replace gallium in GaAs because of the different sizes of the ions. Therefore, it is necessary to assume that GaAs:Mn, like GaAs:Cr, exhibits the Jahn–Teller effect. However, the experimental investigations of the ESR spectra of GaAs:Mn do not provide evidence in support of the existence of this effect but this may be due to the fact that the reported investigations have not been sufficiently thorough and numerous. It should be pointed out that a model postulating the "capture" of a manganese atom by a neutral gallium vacancy is put forward in Ref. 74. A manganese atom transfers two electrons from the 4s-shell vacancy and it is transformed to a $3d^5$ ion. The size of this ion should be affected by the screening action of the electron orbitals of the vacancy because the Mn^{3+} ion in GaAs is not the ion considered in the dimensional analysis given in Sec. 1 of the present chapter.

We cannot exclude the possibility that the two hypotheses – the displacement of a manganese ion relative to the centre of a gallium site and the "capture" of manganese by a gallium vacancy – represent two aspects of the same phenomenon, described in terms of different approaches. The available experimental data are insufficient to judge this topic more thoroughly.

Vanadium

The available information on the vanadium atoms in GaAs is clearly insufficient. It was obtained mainly[75] by studies of the optical and photoelectric properties of GaAs:V single crystals which were strongly compensated and exhibited semi-insulating properties with a resistivity of $\sim 10^8\ \Omega \cdot cm$.

The absorption spectrum recorded at 300 K (Fig. 42) consists of a background in the form of a wide charge-transfer band with a superimposed characteristic (only of this system) structure comprising three relatively wide and overlapping bands (curve 3). Cooling makes these bands assume a resonance profile: two of them with the lower energy increase in intensity and shift considerably towards shorter wavelengths. On the other hand, the intensity of the third band with a higher energy decreases somewhat and its position in the spectrum changes slightly.

Another fine structure but at an energy of ~0.7 eV is exhibited by the photoluminescence spectrum at 6 K (Fig. 43). Since the 4–1, 5–2, and 6–3 energy intervals between the lines are practically the same and correspond to the TO-phonon energy $h\nu = 2$ meV, the peaks 4-6 can be regarded as the phonon replicas of the peaks

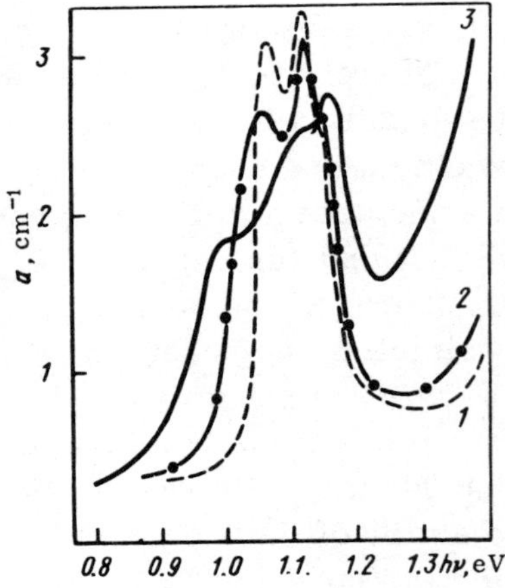

FIG. 42. Absorption spectra of GaAs:V single crystals recorded at 4.2 K (1), 78 K (2), and 300 K (3).

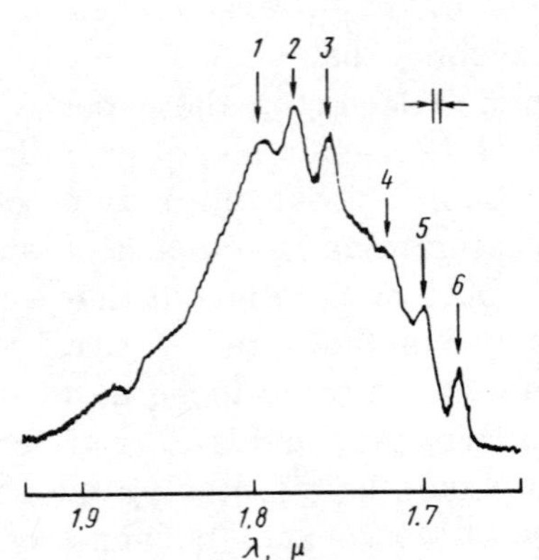

FIG. 43. Photoluminescence spectrum of a GaAs:V crystal at 6 K.

1-3. An increase in temperature destroys the fine structure of the photoluminescence spectrum so that at 77 and 300 K only one fairly strong impurity luminescence band is observed: its energy at the maximum is 0.71 eV and the half-width is 0.06 and 0.13 eV 77 and 300 K, respectively. The position of this band is independent of temperature and this is one of the indications of the intracentre nature of the radiative recombination in GaAs:V in the region of 0.7 eV.

We shall now consider the likely identification of the fine structure of the absorption spectra in the photoluminescence. We shall turn to the scheme of the terms of the $V^{2+}(3d^3)$ ion split by a crystal field of the T_d symmetry (Fig. 11b). According to these data and the selection rules, the infrared absorption spectra should exhibit three spin-allowed transitions from the ground state $^4T_1(^4F)$ to the excited states $^4T_2$, $^4A_2$, and $^4T_1(^4P)$ (Table V).

It is suggested in Ref. 75 that the experimentally observed strongest absorption band at 1-1.2 eV corresponds to the $^4T_1 \rightarrow$ $\rightarrow {}^4T_1(P)$ transitions, whereas the impurity photoluminescence band at 0.7 eV (Fig. 43) is due to the $^4T_2 \rightarrow {}^4T_1$ transitions. The fine structure of the latter band is due to the spin–orbit splitting since the energy intervals 1–2 and 2–3 between the peaks in the photoluminescence spectrum are 58 and 62 cm^{-1}, which is close to the spin–orbit coupling parameter $\lambda = 55\ cm^{-1}$. The expressions in Table V for the $3d^3$ configuration of the V^{2+} ion in the T_d crystal field with the experimental values 1.1 and 0.7 eV substituted in them yield $\Delta = 6100\ cm^{-1}$ and $B = 250\ cm^{-1}$ (Ref. 80). The substitution of these parameters into the expression for another likely transition $^4T_1 \rightarrow {}^4A_2$ (Table V) gives an energy gap of 1.45 eV. The corresponding relatively weak band is difficult to observe against the strong absorption background in this part of the spectrum. These results of investigations of GaAs:V show that vanadium, like the other T-metal impurities from the iron group, is located at the gallium sites in the crystal lattice of gallium arsenide and it forms V^{3+} or V^{2+} ions, depending on the degree of compensation of a sample by accidental or deliberately introduced donors. An earlier report[76] of vanadium in the form of V^{4+} in GaAs has not been confirmed subsequently and may have been due to the use of not very pure (99.99%) vanadium as the dopant.

1.5. STATE OF T-METAL IONS IN GALLIUM PHOSPHIDE

The behaviour of the iron group ions in the crystal lattice of GaP is fully analogous to that reported above for GaAs. However, the experimental investigations of GaP containing T-atom impurities have been fewer and less detailed than those of GaAs.

Applications of the ESR method in studies of the states of T-metal ions in GaP were reported in Refs. 74 and 77-82. The parameters of the impurity ions found in these investigations are listed Table X. Moreover, the state of T-metal ions in GaP has been investigated also by optical spectroscopy.[68, 78, 83-85]

We shall now consider the results of an investigation of the state of the individual T-metal ions in GaP.

Iron

The ESR spectra of GaP:Fe samples always consist of five lines, irrespective of the doping and growth technologies, and these spectra are identical with the spectrum of GaAs:Fe shown in Fig. 29 and representing the iron impurity present in the form of Fe^{3+} with the $3d^5$ electron configuration. This is in agreement also with the results of optical absorption in p-type GaP:Fe with a resistivity of 10^8–$10^9\ \Omega\cdot cm$ (Ref. 78). Such a high resistivity is evidence of strong electrical compensation of the samples, which means that a considerable fraction of the iron ions acquires electrons from the compensating donors and goes over to the Fe^{2+} state with the $3d^6$ configuration. The optical absorption spectrum of such samples exhibits 300 K absorption in the photon energy range 0.4-0.6 eV. This band is of a resonance nature and it is typical of crystals containing the iron impurity in the $3d^6$ configuration. By analogy with GaAs:Fe, this band consists of two superimposed peaks representing intracentre zero-phonon transitions as well as transitions assisted by various phonons and their combinations. The spectrum of this band recorded at low temperatures[78] is shown in Fig. 44, which demonstrates clearly its fine structure. A similar structure was reported for GaP:Fe also in Ref. 50. We analyse this structure in detail in the preceding section when dealing with $GaAs:Fe^{2+}$. Four narrow lines at ν = 3300–3350 cm^{-1} represent intracentre zero-phonon transitions between the ground ^5E and excited $^5T_2$ states, into which the ^5D term of the Fe^{2+} ion splits in the T_d crystal field (Fig. 34). The use of relationships from Table XI (without allowance for the coefficient λ_T representing the Jahn–Teller shift) gives Δ = 3345 cm^{-1} and λ = 82 cm^{-1}.

The spectrum in Fig. 44 has an additional resonance peak at ν = 2648 cm^{-1}. Its intensity[78] is not correlated with the absorption peak of the Fe^{2+} ions and its origin is not clear.

It should be pointed out that, in addition to the normal ESR spectrum, a second spectrum, called the B spectrum, was reported in Ref. 85; it has a completely different g-factor amounting to

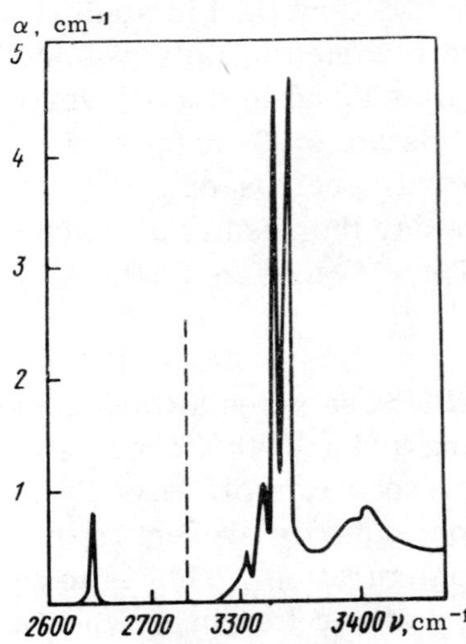

FIG. 44. Absorption spectrum of GaP:Fe at 20 K

2.131 ± 0.002 attributed in Ref. 85 to the $3d^7$ configuration of iron. It was concluded in Ref. 85 that the observed spectrum is not due to a transition of iron atoms from the $3d^5$ to the $3d^7$ state, but represents a different state in which some of the iron atoms are present in a crystal. Both spectra were reported to be very photosensitive.

In other investigations (see, for example, Ref. 84) the additional B spectrum of the ESR was not observed in darkness or during illumination.

It is characteristic to note that the samples used in Ref. 85 were overcompensated with donors (tellurium, tin) and exhibited n-type conduction, whereas all the other investigations have been carried out on compensated samples but with p-type conduction. This circumstance could affect somehow the state of the iron atoms in GaP.

A similar overcompensation of GaAs:Fe:Te (Sec. 1.4) results simply in a complete transfer of iron from the charge configuration $Fe^{3+}(3d^5)$ to the $Fe^{2+}(3d^6)$ configuration and no B spectrum of the ESR is observed.

It seems to us there are no grounds for assuming that iron behaves in any different way in the compounds GaP and GaAs, which are similar in all respects.

Chromium

The chromium impurity in GaP was investigated by optical spectroscopy.[68] A wide resonance absorption peak was found and it was attributed to optical transitions in the Cr^{2+} ion.

Nickel

Unfortunately, no ESR studies have yet been made of GaP:Ni crystals. Therefore, conclusions on the state of nickel atoms in GaP can only be drawn from the results of optical investigations and by analogy with GaAs:Ni.

In accordance with the Tanabe–Sugano diagram for the $3d^8$ configuration of Ni^{2+}, shown in Fig. 11g, we can expect all the transitions manifested in GaAs:Ni:Te in which the nickel ions are completely compensated and are also in the Ni^{2+} state. All these transitions (Table V) were found in GaP:Ni at 4.2 K (Refs. 9 and 86). They were used to determine $\Delta = 5500\ cm^{-1}$ and $B = 310\ cm^{-1}$ (Ref. 9).

In contrast to GaAs:Ni, the absorption spectrum of GaP:Ni (Fig. 45) shows clearly a fine structure due to the additional

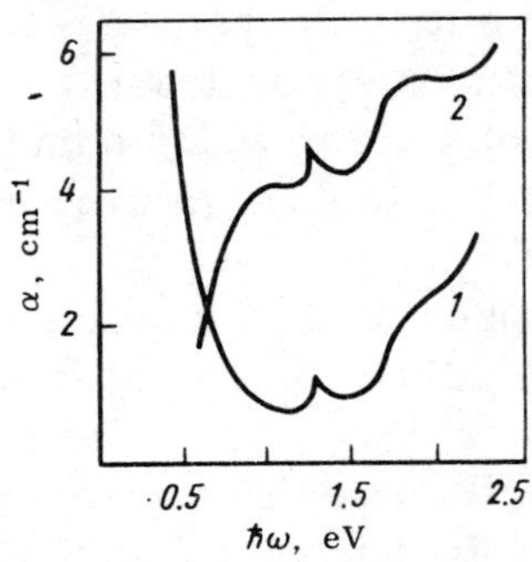

FIG. 45. Absorption spectra of n-type (1) and p-type (2) samples of GaP:Ni at T = 4.2 K.

splitting of the terms because of the spin–orbit interaction. This shows that the excited state $^3T_1(P)$ to which electrons are transferred from the ground state $^3T_1(F)$ is not located inside the conduction band, as is true of GaAs, but inside the band gap. In this case there is no hybridization of the $^3T_1(P)$ impurity state with the continuum and there is no obstacle for the appearance of low-temperature ($kT < \lambda$) fine structure in the optical absorption spectrum.

The value of λ found from the optical data[8] is $-87\ cm^{-1}$ for GaP:Ni.

Manganese

The ESR spectra of GaP:Mn have been observed for both types of conduction[74] and in all cases they exhibit the $3d^5$ state, i.e. manganese is present in the form of the Mn^{2+} ions.

The agreement between the manganese concentration deduced from the ESR spectrum and that found by spectrochemical analysis of the same samples[74] demonstrates that all the manganese centres have the $3d^5$ configuration.

We recall that in the case of GaAs the corresponding spectrum appears only in the presence of external donors. Therefore, the explanation given in Ref. 74, according to which a manganese atom is captured by a neutral vacancy to which it transfers two electrons from the 4s-shell, should be applied to GaP with great caution. There is no convincing proof of the generation of gallium vacancies in the investigated samples.

Unfortunately, no investigations of GaP:Mn have been made

by other methods (such as optical spectroscopy and paramagnetic acoustic resonance), which in combination with ESR and the double doping method (Mn, Te or Mn, Zn) would have made it possible to obtain more reliable information on the state of the manganese impurity in gallium phosphide.

Vanadium

The ESR spectrum of vanadium in GaP was considered in Ref. 82. If the vanadium atoms form, by analogy with iron, chromium, and other T-metal impurities, a substitutional solution in GaP, i.e. if they occupy the Ga sites, the electron structure of these vanadium atoms should be $V^{3+}(3d^2)$. The ground state A_2 of this electron configuration (see Fig. 11a) has zero orbital momentum with the effective spin $S = 1$ and, consequently, we should be able to observe the ESR spectrum of these centres. However, this spectrum was not obtained experimentally for n-type crystals in the presence of a background of shallow donors in a concentration of about 10^{17} cm^{-3} (the authors of Ref. 82 were clearly unable to grow purer GaP:V crystals). In crystals of this kind the whole of the vanadium impurity is ionized in the form of V^{2+} and has the electron configuration $3d^3$. Donor electrons fill the lowest level (Table IV) and a nonzero orbital momentum appears so that the ESR is not observed.

However, there should be a structure in the optical absorption spectrum due to the splitting scheme of the terms of the V^{2+} ion by the crystal field of the T_d symmetry in accordance with the Tanabe–Sugano diagram (Fig. 11b). Such a structure has indeed been found in the absorption spectrum of GaP:V (Ref. 83) and it was attributed to electron transitions within the d-shell of the V^{2+} ion, fully analogous with the behaviour of GaAs:V discussed in Sec. 1.4.

The type of conduction of GaP:V samples can be altered to p-type by the diffusion of both vanadium and manganese into the original gallium phosphide.[82] These samples exhibit an ESR spectrum of manganese ions, but additionally an ESR spectrum of vanadium ions in the $3d^2$ state, i.e. of the V^{3+} ions, was observed. The spectrum consisted of two fine-structure lines. The parameters of this spectrum are listed in Table X.

Since the spin of the V^{51} nucleus is 7/2 and its natural abundance is 99.76% (Table VI), the spectrum should exhibit eight hyperfine-structure lines, but they are not observed experimentally and this is attributed in Ref. 82 to the broadening of the hyperfine-structure lines because of the superhyperfine interaction of the d-electrons of the V^{3+} ion with the nucleus of this ion.

1.6. STATE OF T-METAL IONS IN OTHER III-V SEMICONDUCTORS

Indium Phosphide

Information on the state and behaviour of T-metal impurities in InP crystals is scarce. However, a determination of the ESR and optical absorption spectra at helium temperatures carried out on InP:Fe (Ref. 87) led to the same conclusions as in the case of GaAs and GaP. In fact, it is clear from Fig. 46 that in the region of 0.3-0.6 eV the absorption of light may be attributed to optical transitions of electrons within the Fe^{2+} ions. This is manifested clearly when samples are cooled to 20 K. It is reported in Ref. 87 that the band then splits into four narrow lines typical of the transitions between the states ^5E and $^5T_2$ of the Fe^{2+} ion with the $3d^6$ configuration split by the T_d field of the indium phosphide lattice. These lines yield the splitting parameters $\Delta = 3025\ cm^{-1}$ and $\lambda = -87\ cm^{-1}$. The ESR spectrum determined at 100 K at the frequency of 9.3 GHz (Ref. 87) consists of five anisotropic lines of width 120 Oe; the spin is $S = 5/2$, and also $g = 2.02$ and $a = 223 \cdot 10^{-4}\ cm^{-1}$. These data are in agreement with those listed in Table X and they confirm that iron in InP can have both the $Fe^{2+}(3d^6)$ and $Fe^{3+}(3d^5)$ charge states; the ratio of these concentrations in both states is governed by the degree of compensation of the deep Fe^{3+} centre.

The state of the manganese ions in InP was investigated using samples grown by the Czochralski method and epitaxial $Ga_xIn_{1-x}P$ solid-solution films.[88] When x was varied from 1 to 0, the parameters of the ESR spectra ranged from the values typical of GaP:Mn to those corresponding to InP:Mn. In the latter case the ESR spectrum consisted of a single line. Its parameters are listed in Table X. The form of the spectrum can be represented as a superposition of six hyperfine-structure lines[88] of the $Mn^{2+}(3d^5)$ centre located at the indium lattice site. It is suggested in Ref. 88 that an atom of manganese $(3d^54s^2)$ contributes only two electrons to the bonds, in contrast to three electrons in the case of iron. The d-shell is not affected. However, it seems to us that it is more realistic to expect the same situation as in other III-V systems containing T-metal ions: three electrons are used

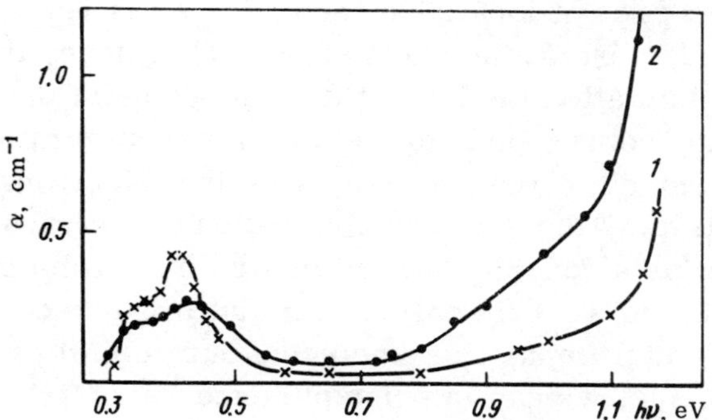

FIG. 46. Absorption spectra of InP:Fe samples obtained at T = 20 K (Ref. 87). The concentration of the Fe^{3+} ions for the sample represented by curve 1 was higher than that for the sample corresponding to curve 2.

to form the bonds and the ground nonionized state is $Mn^{3+}(3d^4)$, but the background of donor impurities present in a crystal in amounts exceeding the manganese concentration results in the ionization of manganese so that it is in the $Mn^{2+}(3d^5)$ state. This can be checked by paramagnetic acoustic resonance and optical spectroscopy using similar samples as well as samples containing additionally introduced shallow acceptors (zinc) for the purpose of neutralization of the donor background.

Indium Arsenide

Electron spin resonance in InAs:Fe reveals[89] the $3d^5$ state, i.e. the Fe^{3+} ions occupy the indium sites. The parameters of these ions[89,90] are given in Table X. Unfortunately, these two investigations are the only ones that give information on the state of iron impurities in InAs.

We shall now consider an investigation of the state of chromium in InAs (Ref. 91). It was reported in that paper that introduction of chromium into pure crystals with a low background of residual donors ($N_d^* \sim 2.10^{16}$ cm^{-3}) did not alter the electrical properties of the samples. When chromium was introduced simultaneously with a shallow acceptor (zinc) in amounts so that the conditions $N_{Zn} > N_d^*$ and $N_{Cr} > N_{Zn} - N_d^*$ were obeyed, these crystals were found to have a high resistivity and the electron density at 77 K was 10^{11}-10^{12} cm^{-3}, which was more than four orders of magnitude less than in the undoped material. This result suggests that chromium in InAs behaves as an acceptor rather than as a donor. We shall return to this topic in Chap. 3. If we assume that the chromium impurity atoms form substitutional solid solutions, then they are not located at the cation sites (as in wide-gap III-V compounds), but at the anion sites. Therefore, the neutral state of chromium after the loss of five electrons in the formation of valence bonds corresponds to the electron configuration $3d^1$. The Cr^{5+} ion can quite readily give up its last electron from the 3d-shell and assume the stable configuration of the rare gas (argon), i.e. it may act as a donor. It is not clear why chromium replaces the metalloid in InAs. It is only known that in some compounds the transition metal impurities (including chromium) have a valence in excess of 3, i.e. we can in principle expect the Cr^{5+} and Cr^{6+} ions.

It is interesting to note that the size factor (Sec. 1.1) also forbids the occupancy of the indium sites and it also forbids the occupancy of the As sites (Table II).

All our attempts to determine experimentally the local symmetry of the chromium impurity centres in InAs by the ESR and PAR methods were unsuccessful. The paramagnetic absorption signal was not observed.

Indium Antimonide

As in the case of indium arsenide, the chromium impurity introduced into InSb simultaneously with a shallow acceptor (zinc) reduces greatly the density of free electrons[91] compared with undoped crystals. Therefore, in the case of InSb:Cr we can also assume that chromium replaces the antimony atoms and it is in the non-ionized state $Cr^{5+}(3d^1)$ or in the ionized state $Cr^{6+}(3d^0)$. This hypothesis has not yet been proved because attempts to observe ESR and PAR signals in InSb:Cr and InSb:Cr:Zn crystals have not been successful and a reliable determination of the donor nature of the chromium impurity is also difficult; moreover, it is disputed in Ref. 92 and the topic will be discussed in greater detail in Chap. 3.

The examples of InAs and InSb show that the behaviour of the T-metal ions belonging to the iron group differs considerably from the behaviour of the same ions in III-V semiconductors with wider energy gaps. It should be pointed out that this applies not only to transition elements. For example, the copper impurity is a typical acceptor in the majority of semiconductors, with the exception of InAs and InSb, in which it acts as a donor. Therefore, the final conclusions on the state of chromium in the crystal lattices of narrow-gap III-V semiconductors cannot be drawn without further investigations.

1.7. ASSOCIATED STATES OF T-METAL IONS IN III-V SEMICONDUCTORS

Under suitable conditions a T-metal impurity in a III-V crystal may interact with the other impurities and form complex associated defects. A necessary condition for the appearance of such defects is a rapid migration of at least one of the partners of the interaction. Moreover, certain forces favouring the interaction should act in the crystal. The most frequently encountered is the simple electrostatic Coulomb interaction of a T-metal ion and an oppositely charged atom of another impurity located at a short distance within one or two coordination spheres. Associates with larger internal distances can also form if the second component is a hydrogen-like impurity with the Bohr radius extending over several lattice periods because of a small effective electron mass. The effective mass at the bottom of the conduction band of GaAs is only $0.07m_0$ and the Bohr radius extends to 8.5 nm in the lattice with the period of 0.565 nm. Therefore, the Coulomb interaction between T-metal ions acting as acceptors in GaAs with hydrogen-like donors forming shallow levels is quite probable.

Nevertheless, formation of impurity pairs of any T-metal ions with, for example, tellurium has not been confirmed even in the case of samples with deliberate double doping, such as GaAs:Fe:Te, GaAs:Cr:Te, and GaAs:Ni:Te. Similar experiments but with tellurium replaced with selenium have not yet been carried out, but if the shallow impurity is sulphur, then associated defects of the Mn^{2+}–S type can form, as indicated by the ESR spectra.[81,93,94]

The formation of associates with sulphur atoms but not with tellurium atoms suggests that the binding of atoms in an associate is more "chemical" rather than due to the simple Coulomb interaction, i.e. the associates are composed only of those impurity atoms which are capable of forming chemical compounds (FeS, MnS, NiS) and not of those that do not form compounds.

Naturally, this principle should be checked by considering the results of investigations of the ESR spectra in GaAs doped with sulphur and also with iron, nickel, cobalt, chromium, or vanadium, because the effects of doping of GaAs with these metals alone have been investigated quite thoroughly, as demonstrated in the preceding sections.

The results of investigations of GaAs:Mn:Li crystals were reported in Refs. 85, 93, and 94, where the ESR spectra indicated the presence of Mn^{2+}–Li complexes. However, in this case the complexes were formed purely by the Coulomb interaction if the composition of the complex was as indicated. A characteristic feature of these and other complexes is a lowering of the symmetry of a complex compared with that of a single T-metal ion in GaAs. This is evidence of some preferred directions in a crystal along which association is more likely. For example, it was reported in Ref. 93 that the Mn^{2+}–S centres have an axial symmetry along the [111] crystallographic direction. An important form of an associated defect results from the interaction between a T-metal ion and a vacancy. This interaction was observed in epitaxial GaAs films doped with manganese.[94] These films were grown by the method of liquid phase epitaxy. The manganese impurity was added to molten Ga. A study was made of the angular dependence of the fine-structure lines in the ESR spectrum. This structure could be described satisfactorily by postulating the formation of Mn_{Ga}–V_{As} complexes. In principle, it could also be explained by postulating that some other impurity was present instead of the vacancy. However, it would be difficult to see how a crystal can capture $\sim 3 \cdot 10^{19}$ cm^{-3} of a second impurity, as this would be the concentration required to account for the presence of $3 \cdot 10^{19}$ cm^{-3} paramagnetic centres. On the other hand, the concentration of the arsenic vacancies in GaAs could be high because of deviations from stoichiometry. It should be pointed out that these conclusions should be checked very carefully in the quantitative sense because at the relatively low growth temperature of epitaxial films one could hardly expect this concentration of vacancies.

The absence of similar centres in bulk gallium arsenide may be explained by the higher growth temperature of such crystals when dissociation rather than association of T-metal ions and vacancies prevails. It is possible that doping by diffusion at various temperatures will provide information on this topic.

Specific interactions have been observed as a result of heavy doping of semiconductors with T-metal impurities. For example, if the concentration of iron is close to the solubility limit or even slightly above it, it has been found on several occasions[95] that the iron atoms interact in the $3d^5$ state. This interaction is manifested clearly in the ESR spectrum shown in Fig. 47. In addition to the fine-structure lines of the single (isolated) Fe^{3+} centres, the spectrum includes also lines with a total spin of $S = 5$ which are re-

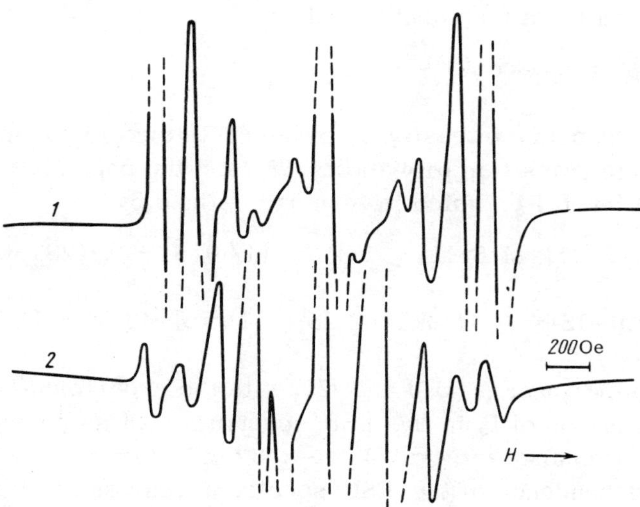

FIG. 47. Electron spin resonance spectra of exchange-coupled pairs in GaAs:Fe. Magnetic field orientation: 1) H ‖ [011]; 2) H ‖ [001].

garded in Ref. 95 as due to exchange-coupled pairs of iron atoms.

It should be pointed out that such a spectrum can be observed only if a sample contains $(3\text{-}5)\cdot 10^{18}\ \mathrm{cm}^{-3}$ of the iron atoms without significant amounts of inclusions of the second phase, which is difficult because this amount of the impurity is somewhat higher than the solubility limit of iron in GaAs under equilibrium conditions. Consequently, samples of this kind are naturally metastable. Moreover, the ESR spectrum of the exchange-coupled states of the iron atoms in GaAs can be observed only at sufficiently high frequencies, because at low frequencies and in the weak magnetic fields the spectrum cannot be resolved. In fact, it was pointed out in Ref. 95 that when the observation frequency was reduced from 28 to 10.5 GHz, the spectrum shown in Fig. 47 transformed into a single broad line.

It may be assumed that two nearest iron atoms occupying sites in the gallium sublattice are coupled along the [011] axis, which can be regarded as the z axis. Then, in the case of a strong exchange interaction the spin Hamiltonian of a pair can be written in the form[96]:

$$H = g\beta\overrightarrow{HS} + D(S^2_{Ax} - S^2_{Bx}) + \frac{B}{2}[(\vec{S}_A\vec{S}_B) - 3S_{Az}S_{Bz}], \qquad (17)$$

where $S = S_A + S_B$; g and D are the parameters of the spin Hamiltonian for a free centre and

$$B = (g^2\beta^2/r^2_{AB})(3\cos^2\theta - 1). \quad (18)$$

The average intensity I_S of an ESR line is proportional to the average transition probability W_S and the population of the exchange level P_S, which in turn are defined by

$$W_S = (1/2^S)\sum_M |\langle M|S|M + 1\rangle|^2 = (1/6)(S + 1)(2S + 1); \quad (19)$$

$$P_S = \exp[-JS(S + 1)/2kT]/\sum_S (2S + 1)\exp[-JS(S + 1)/2kT]. \quad (20)$$

Comparing Eqs. (19) and (20) with the experimental results of a determination of I_S in the ESR spectrum, it was possible to find[95] the exchange integral $J = -8.3\ \mathrm{cm}^{-1}$ for $S = 5$. The angular dependence of the ESR spectrum was used in Ref. 95 to determine the constant B and then, using Eq. (18), to find the average distance between the atoms in a pair amounting to 0.49 nm, which compared with the GaAs lattice period ($a = 0.565$ nm) led to the conclusion that the pair had the $Fe(3d^5)$–As–$Fe(3d^5)$ structure.

A further increase in the concentration of iron in GaAs to $10^{20}\ \mathrm{cm}^{-3}$ resulted in a stronger interaction between the pairs themselves giving rise to superparamagnetic regions representing inclusions in the diamagnetic matrix of gallium arsenide. Superparamagnetism of GaAs:Fe will be considered in greater detail in Chap. 4.

The interaction of iron ions is possible naturally also in those cases when they are present in the form of Fe^{2+} with the electron configuration $3d^6$. Formation of Fe^{2+} pairs gives rise to an optical absorption band in the spectrum shown in Fig. 33: the band is located at $\nu = 3161\ \mathrm{cm}^{-1}$ and it is exhibited only by the samples with the highest iron concentrations.

1.8. MODEL OF THE STATE OF T-METAL IONS IN CRYSTAL LATTICES OF III-V SEMICONDUCTORS

The results of investigations of T-metal ions in III-V semiconductors discussed in the preceding sections have made it possible to establish quite reliably the main features of the state of these ions in crystal lattices, at least in wide-gap semiconductors such as

GaP, GaAs, and InP. In all these lattices a T-metal impurity is in the form of a substitutional solution and it occupies the sites in the A^{III} sublattice. This is the main difference between the behaviour of T-metal impurities in III-V compounds from their behaviour in germanium and silicon.

The T-metal atoms occupying the A^{III} sites participate in the formation of chemical binding with atoms in the B^{V} sublattice and give up three electrons from the $3d^{n}4s^{2}$ configuration for the formation of coordination bonds. Therefore, in the nonionized state a T-metal impurity has the valence of 3, i.e. it is present in the form of Me^{3+} with the $3d^{n-1}$ configuration. This is completely analogous with the behaviour of T-metal atoms in silicon and germanium in those cases when these atoms occupy the site positions in these elements. Therefore, in the case of III-V crystals we are even more justified (than in the case of silicon or germanium) in the application of the Roĭtsin–Firshteĭn–Ludwig–Woodbury (RFLW) model with its conclusions that the wave functions are localized in the second coordination sphere and the split multiplets are filled with electrons in accordance with the Hund rule.

Experimental investigations have demonstrated splitting of the terms of the T-metal ions in III-V semiconductors in complete qualitative agreement with the crystal field theory. Naturally, major changes in the main parameters of this theory are expected in the case of III-V compounds: the value of B should decrease strongly and Δ should rise greatly compared with the corresponding parameters of the compounds with stronger ionicity, including II-VI semiconductors, as demonstrated clearly in Table XII in the case of some of the Me^{2+} ions. If the measure of covalence is the electron permittivity $\varepsilon_e \sim n^2$, where n is the optical refractive index, the experimental values of B for the T-metal impurities in III-V compounds can also be described by the empirical dependence (5) established earlier for II-VI compounds with T-metal impurities. The second parameter of the crystal field theory – the quantity Δ – behaves in a more complex manner. It is pointed out in Sec. 1.2 that the relationship (5) between Δ and the degree of ionicity on the Phillips scale applies also to II-VI compounds.[9]

This quantitative relationship is fairly unexpected. It is reasonable to assume that introduction of a foreign impurity atom into the lattice should cause a local deformation, redistribute the electron charge density, alter the force constants and the polarizability, and produce other effects. In particular, the values of f

TABLE XII. Values of Energy Δ of T-Metal Ions in III-V and II-VI Crystals with Different Ionicities f

f	Crystal	Δ, cm^{-1}, of T-metal ions					
		V	Cr	Mn	Fe	Co	Ni
0.31	GaAs	4500	—	—	2995	4000	5500
0.327	GaP	5400	—	—	3345	5400	5600
0.421	InP	—	—	—	3025	—	—
0.623	ZnS	—	—	5000	3300	3600	4750
0.63	ZnSe	3200	5500	4000	3100	3750	4400
0.68	CdS	4350	5200	—	3150	3200	4000
0.699	CdSe	3800	—	—	2400	—	—
0.717	CdTe	3560	—	—	2500	3100	—

Note. There are no experimental data on scandium in III-V and II-VI semiconductors. In the case of titanium in ZnSe the value Δ = 3700 cm^{-1} has been reported.

near an impurity ion should deviate considerably from the ionicity of the regular lattice. On the other hand, it is quite clear that the energy spectrum of the impurity ions is governed primarily by the immediate environment. Therefore, there are some doubts about the validity of using, in quantitative comparisons, the relationship (5) linking the ionicity of an ideal crystal with Δ.

The values of Δ for T-metal ions in various III-V crystals (Table XII) do not fit the linear dependence of Eq. (5). Therefore, the observation that such a dependence is obeyed by Ni^{2+} and Co^{2+} in II-VI compounds should be regarded as purely accidental.

There is another correlation which is manifested clearly by the data in Table XII: the value of Δ for each of the III-V semiconductors rises in the transition metal series as the occupancy of the d-shell is increased above the half-filled state (Fe^{2+}, Co^{2+}, Ni^{2+}). In the case of the T-metal ions with the d-shell less than half-filled (V^{2+}, Cr^{2+}), the dependence is just the opposite. This is true also of II-VI semiconductors. The d^5 configuration of Mn^{2+} forms a kind of "watershed". It is interesting to note that on both sides of manganese there are opposite dependences of the ionic radii in T-metal series, as demonstrated clearly in Figs. 5-7.

Investigations of the ESR spectra of T-metal ions in the non-ionized Me^{3+} state have made it possible to determine the phenomenological constant of the Hamiltonian (8) listed in Table X. However, the absence of a quantitative theory of the ^6S state makes it impossible to draw any definite conclusions on the nature of the binding of a paramagnetic ion and its environment, on the spatial distribution and size of the impurity wave function, etc. We may hope that

TABLE XIII. Parameters of ESR Spectra of Fe^{3+} Ions in III-V and II-VI Compounds of Tetrahedral Symmetry

Crystal	Δg	$a \cdot 10^4$, cm^{-1}	ΔH, Oe	$\Delta H/\beta_N^B$	$\Delta H/\beta_N^A$
GaAs	0.045	340	53	37	24.2
InAs	0.033	420	130	91	23.6
GaP	0.023	390	40	44.2	22.8
InP	0.018	220	120	106	21.8
ZnS	0.017	128	0,7	60.5	15.3

at least partial success in this direction might be achieved when the results are compared with the same iron impurity in different crystal matrices. With this in mind we shall consider the data in Table XIII, which gives the results of the ESR measurements carried out on the Fe^{3+} ions in the series of III-V and II-VI compounds.

We can see that in the case of all the substances with the exception of GaAs a reduction of the shift of the g-factor compared with the g-factor of a free electron, $\Delta g = g - 2.0023$, is accompanied by a reduction in the cubic splitting parameter a. The observed change in these quantities can be attributed to an increase in the ionicity or to the corresponding reduction in the covalence of the binding in this series of compounds. In fact, according to the well-known calculations of Watanabe,[96] a positive value of $\Delta g > 0$ for the ions in the ^6S state can be explained only by the presence of a covalent component in the impurity–matrix chemical binding. It should be pointed out that similar changes in the parameters of the ESR spectra of ions in the ^6S state are exhibited by Cr^+ in zinc chalcogenides,[97] and by Mn^{2+} in lead[41] and zinc[97] chalcogenides.

As already pointed out, the absence of a theory makes it impossible to give a quantitative treatment of changes in Δg and a in various crystals. It is only known that these two parameters are governed by the constant representing the splitting by the crystal field Δ and the spin–orbit interaction λ, which in turn may vary with the degree of ionicity of the binding. Figure 48 shows, on a double logarithmic scale, the dependence of a on Δg, which in the first approximation can be regarded as a straight line with the slope ~1.3. It follows that the value of a depends more strongly on Δ and λ than does Δg.

It is worth noting the sequence of the compounds in Table XIII

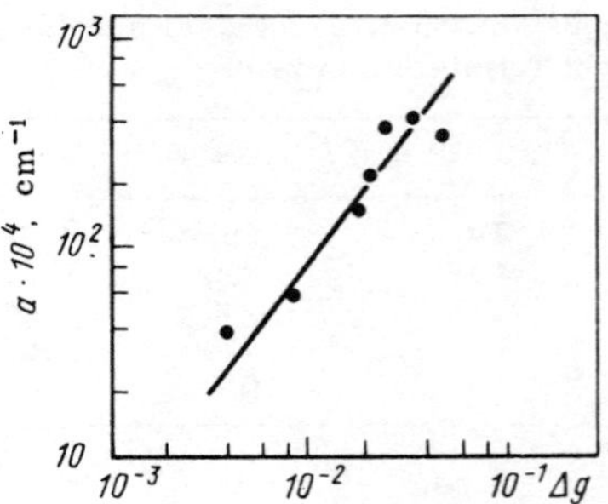

FIG. 48. Relationship between the cubic splitting constant a and the shift of the g-factor in semiconductor compounds of the T_d symmetry.

arranged in the decreasing order of Δg. The compounds with the same B^V element are arranged in pairs, because in these pairs the sequence of the A^{III} elements is retained. It follows that changes in the values of Δg, a, and clearly also the ionicity are affected primarily by the immediate environment, whereas the influence of the A^{III} atoms is of less importance. It is interesting to consider the variation of the width ΔH of an ESR line. We can see that in the III-V and II-VI series of compounds there is no definite relationship for ΔH. The most probable reason for the broadening of the ESR resonance lines of the Fe^{3+} ion in the ^6S state is the superhyperfine interaction with the ligand nuclei of intensity proportional to the magnetic moments of these nuclei. One can therefore expect that an increase in the ionicity of the binding should reduce the superhyperfine interaction. Table XIII lists not only ΔH but also the calculated values of $\Delta H/\beta_N^B$ and $\Delta H/\beta_N^A$, where β_N^B and β_N^A are the magnetic moments of the nuclei of the elements in the first and second coordination spheres around a Fe^{3+} ion. An analysis of these values leads to the conclusion that the width of the Fe^{3+} ESR lines is mainly due to the superhyperfine interaction not with the anion (nearest) but with the cation (second-nearest) environment of the paramagnetic ion. It is interesting to note also that the intensity of the magnetic field created by the Fe^{3+} ion at the positions of the cations is practically the same for all the III-V compounds. On the other hand, for reasons which are not clear, the superhyperfine interaction with the B atoms in the first coordination sphere is weak and it has little effect on ΔH. It should be pointed out that similar conclusions have been reached also by the investigators (for a bibliography see

Ref. 97) of the Cr^{+} and Mn^{2+} ions in the II-VI series of compounds. The constancy of the magnetic field intensity at the second ligands is explained by the fact that an increase in the covalence of the system is compensated by an increase in the distance between the impurity ion and the nuclei of the nearest cations. The value of $\Delta H/\beta^{A}$ for the Fe^{3+} ion in all crystals of the T_d symmetry decreases slowly but monotonically on increase in the bond ionicity. In III-V compounds, the ionicity of which does not vary greatly, the wave function of Fe^{3+} is fairly large and it ensures the superhyperfine interaction with the nuclei of ions in the second coordination sphere. In the case of ZnS which has the cubic structure the value of $\Delta H/\beta^{A}$ falls by just 30%; clearly, the nature of the superhyperfine interaction is still the same in this compound.

It is usual to assume that the wave functions ψ of deep impurity states are strongly localized. However, the magnetic resonance data reported above clearly indicate that the wave functions ψ in the investigated crystals are quite large. In the final analysis the problem of the size of a wave function is quantitative. This shows how important it is to obtain quantitative information on the electron density distribution for the development of the fundamental ideas on T-metal ions in III-V semiconductors; the necessary investigations of the electron density will have to be carried out by the ENDOR method in systems comprising a III-V semiconductor with a T-metal impurity, and it will be necessary to develop a theory providing a satisfactory description of the state of T-metal impurities in the crystal lattices of these semiconductors.

The results of experimental investigations of the ESR spectra provide evidence of a considerable redistribution of the charge density between the impurity atoms and the atoms in the immediate environment right up to the stage when its maximum is located in the second coordination sphere. The experimental data on optical spectroscopy are described well by the crystal field theory (though with modified parameters Δ and B) which postulates a strong localization of the impurity electrons within the limits of the central T-metal ion.

A self-consistent theoretical description of these contradictory properties of T-metal impurities in semiconductors is the stumbling block in the development of a suitable theory. Various theoretical approaches are discussed in detail in Chap. 2. We shall mention here the attempt of Il'in and Masterov[95] to solve the fundamental contradiction mentioned above by introducing the concept of an impurity quasi-atom with the electron states which are linear

TABLE XIV. Values of Δ for GaAs and GaP Containing T-Metal Impurities

T-metal centre	Δ, eV, for semiconductors	
	GaAs	GaP
Ti	0.86/–	0.66/–
V	0.74/0.756	0.49/0.669
Cr	0.54/–	0.36/–
Mn	0.43/–	0.31/–
Fe	0.35/0.371	0.29/0.414
Co	0.33/0.496	0.27/0.669
Ni	0.40/0.682	0.14/0.694

Note. The numerator gives the calculated values and the denominator gives the experimental results.

combinations of the wave functions of the T-impurity atom and the atoms of the environment.

Il'in and Masterov proposed to obtain a one-electron spectrum of such a quasi-atom by the method of calculation of the electron spectra of disordered systems proposed by Heine et al.[98]

In contrast to the pure cluster approach, in this method a one-electron Green function represented by a continued fraction allows not only for the properties of the immediate environment but, also for those of the whole system. This makes it possible to calculate the total density of states of the system and to identify the edges of the allowed electron energy bands.

This method was used in Ref. 95 to determine the positions of the electron levels of the T-metal centres relative to the allowed band edges (this will be discussed in Chap. 3) and to calculate the charge density of these levels. In view of the presence of a large number of the d-electrons, it was necessary to allow for many-electron effects which modify the one-electron spectrum of a T-metal centre in a crystal. The concept of an impurity quasi-atom makes it possible, according to Ref. 95, to allow for many-electron effects and to calculate the total spin of the ground state and the value of the Δ splitting parameter, i.e. the energy gap between the ground and the first lowest excited state of a T-metal centre. The accuracy of the results obtained can be judged by comparing the calculated and experimental values of Δ given in Table XIV. We can see that the theory does not provide a good agreement with the experimental results. It should be pointed out that the values of the band gap 1.78 eV for GaAs and 2.12 eV for GaP deduced from this theory also differ from the experimental values 1.53 eV and 2.28 eV, which is clear evidence of shortcomings of the theory in spite of the undoubted attractiveness of the quasi-atom concept.

It is possible that a more rigorous allowance for many-electron effects will in future provide a better agreement between the quasi-atomic theory and experiments. At this stage we can simply repeat what we have said earlier: the best theoretical explanation of the state of T-metal impurities in semiconductors is currently provided by the crystal field theory in its semi-empirical variant when the fitting parameters Δ and B are found from the experimental results.

2. ENERGY SPECTRUM OF DEEP LEVELS OF IMPURITY CENTRES THEORETICAL REPRESENTATIONS

2.1. EFFECTIVE MASS METHOD

It is usual to assume that, in principle, the problem of local states in semiconductors was solved about 12 years ago by the effective mass method (EMM) developed mainly by Kohn and Luttinger.[99,100] They were able to show that the concept of an effective mass, which characterizes the dispersion law of carriers in the allowed bands of an ideal crystal $E(k)$, is suitable for the description of the state of an electron in a nonperiodic force field created by various defects of a crystal, including impurity centres. Two fairly general assumptions – smooth variation of the potential of a defect perturbing the periodic potential of a crystal and smallness of the binding energy (ionization energy) of a charge carrier compared with the band gap $E_i \ll E_G$ – reduces determination of the energy spectrum of an impurity and of the corresponding wave functions to a solution of the relatively simple one-electron problem described by the equation

$$[-(\hbar^2/2m)\Delta + V(\vec{r})]F(\vec{r}) = EF(\vec{r}), \tag{21}$$

where $F(\vec{r})$ is known as the envelope function and it is related to the required wave function of the problem $\psi(\vec{r})$ by the relationship $\psi(\vec{r}) = F(\vec{r})U_{n,0}(\vec{r})$. Here, $U_{n,0}(\vec{r})$ is the normalized Bloch function near an extremum of the conduction band where $k = 0$.

The condition of smoothness of the impurity potential can be expressed in the form

$$a_0 = |\nabla V|/|V| \ll 1, \tag{22}$$

where a_0 is the lattice constant.

If the potential energy $V(\vec{r})$ of a carrier localized in the field of an impurity centre is represented by the energy of the Coulomb interaction allowing for the static permittivity of a crystal, $V(r) = -e^2/\varkappa r$, Eq. (21) reduces to the wave equation for the hydrogen atom and its solutions are

$$E = -Ry^*/n^2, \tag{23}$$

where Ry* is the energy of the ground state of the impurity in the hydrogen-like approximation (the effective Rydberg is $Ry^* = e^4m/2\varkappa^2\hbar^2$, where n = 1, 2, ...).

The envelope function of the ground state is

$$F(\vec{r}) = [1/(\pi a^3)^{\frac{1}{2}}]\exp(-r/a), \tag{24}$$

where the radius of the orbit of the impurity electron is $a = \hbar^2\varkappa/me^2$, the quantity a being much smaller than the lattice constant.

The above simple relationships for the energy and characteristic size of the wave functions are obtained on the assumption that the dispersion law E(k) is isotropic, which is true of many III-V, and II-VI compounds with the absolute minimum of the conduction band at the centre of the Brillouin zone where k = 0.

The results taken from Ref. 101 and presented in Table XV demonstrate conclusively that the simplest variant of the EMM for shallow donors in some crystals describes satisfactorily the experimental results. The agreement is particularly good in the case of excited impurity states.

In the case of elemental semiconductors such as germanium and silicon, which have a complex structure of the conduction band, the ionization energies of the 1s-levels* of group V donors calculated from Eq. (23) differ much more from the experimental values than in the case of semiconductors. Nevertheless, a comparison of the theory with experiment shows that the simplest variant of the EMM can describe correctly (to within one order of magnitude) the energy of the ground s-type state of an impurity and gives values of the ionization energy of excited p-states of hydrogen-like centres which are very close to the experimental values.

In the case of direct-gap semiconductors of the GaAs type the ionization energies of group VI and IV donors are similar (Table XV) and differ relatively little (by no more than 10%) from the calculated values.

However, the EMM cannot account for the dependence of the ground-state energy on the chemical nature of the dopant. For example, in the case of germanium and silicon the differences between the values of E_i for group V donors reach 30 and 100%, respectively.

*The similarity of the energy spectrum of shallow impurities and of the hydrogen atom has led to the borrowing of the notation system for the ground and excited states 1s, 2s, 2p, etc.

TABLE XV. Ionization Energies of Shallow Donors in Some Semiconductors

Substance	$\varkappa$	$\frac{m}{m_0}$	E_{1s}, MeV		E_{2p}, MeV
			theory	experiment	theory
GaAs	12.53	0.066	5.67	6.08 (Ge); 5.81 (Si); 5.89 (Se); 6.1 (S); 5.87 (Sn)	1.42*
InP	12.60	0.080	6.80	7.28	1.70
CdTe	10.00	0.096	12.96	13.78	3.24
CdSe	9.00	0.110	18.33	–	–

Note. The symbols of dopants in GaAs are shown in parentheses.

* The experimental value is E_{2p} = 1.44 for GaAs.

The EMM is completely unsuitable either for the description of the energy spectrum of isoelectronic and deep impurity centres.

In the 25 years since the work of Kohn and Luttinger, the EMM has been developed further. This has been done in various ways, namely by refinement of the impurity potential near a defect,[102-104] allowance for the spatial dependence of the dielectric screening of the impurity field $\varkappa(r)$, which is important in the vicinity of an impurity centre,[104-107] and allowance for the real energy band structure characterized by the presence of equivalent and additional conduction band extrema.[108-111]

It is appropriate to mention here that the authors of the majority of such investigations have made calculations in which only one factor is allowed for and the influence of the other factors equally important have been ignored. Naturally, all this has reduced the value of the theoretical results obtained in the sense of their quantitative use in the experiments, but it has provided evidence of the great complexity of the problem as a whole.

A detailed analysis of all the refinements to the EMM can be found in Refs. 101-112. Therefore, we shall briefly give only some of the results that reflect the progress made in the EMM as used for calculating the energy spectra of shallow local states. This is best done in the case of elemental semiconductors germanium and silicon, for which the spectrum of shallow donors and acceptors is best known.

In both these substances the effective electron mass is anisotropic and the absolute minima are located along the [111] axis for germanium and [100] for silicon, and the constant-energy surfaces represent ellipsoids of revolution, governed by the familiar

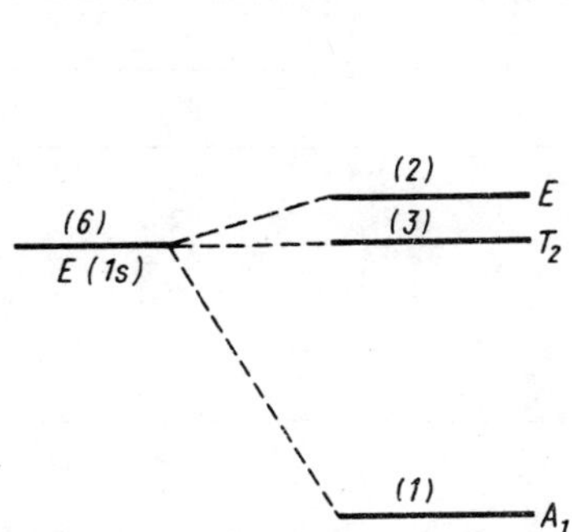

FIG. 49. Splitting of the ground state of a donor in silicon. The numbers in parentheses show the multiplicity of the level of degeneracy without allowance for the spin degeneracy.

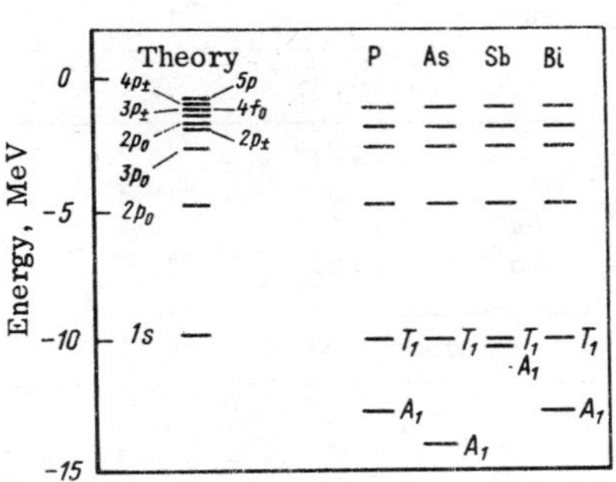

FIG. 50. Energy spectrum of donors in germanium.[113]

values of the longitudinal and transverse effective masses. The number of such equivalent ellipsoids N is set by the cubic symmetry of the crystals and it amounts to 4 and 6 for germanium and silicon, respectively, so that the local states are N-fold degenerate. In this case the wave function of an impurity electron is described by

$$\psi(\vec{r}) = \sum_{j=1}^{N} \alpha_j F_j(\vec{r}) U_j(\vec{r}),$$

where the coefficients α_j are governed by the symmetry of the impurity states.

It is important to note that the deviations of the impurity potential from the pure Coulomb form in the direct vicinity of a centre results in splitting of the impurity states, the nature of which can be deduced simply from the symmetry considerations. For example, the ground six-fold degenerate state of a donor in silicon splits into a singlet (A_1), a triplet (T_2), and a doublet (E); this splitting is shown in Fig. 49. In the case of germanium the ground state of a donor splits into two, one of which is a singlet and the other a triplet.

Baldereschi[109] demonstrated the need to allow for the intervalley mixing effect or the intervalley interaction which results in splitting. He allowed for the dependence of the permittivity on the wave vector, particularly important at high values.[106] The results

TABLE XVI. Binding Energy of Split Levels of Ground 1s State of Donors in Germanium and Silicon (MeV)

Donor	Ge		Si		
	A_1	T_2	A_1	T_2	E
–	9.81*	9.21*	31.27*	20.67*	19.57*
P	12.90	9.9	45.5	33.9	32.6
As	14.17	10.0	57.7	32.6	31.2
Sb	10.32	10.0	42.7	32.9	30.6

*Theoretical values.

of Baldereschi's calculations together with the known experimental data on germanium and silicon are given in Table XVI.

The energy spectrum of the excited 2s, 2p, 3s, etc. states of donors in germanium and silicon was calculated by Faulkner[113] by a variational method allowing for the effective mass anisotropy. The results obtained are presented in Figs. 50 and 51. We can see that in contrast to the ground state, the EMM provides a fully satisfactory description of the spectrum of the excited p-states of donors. In fact, in this case the influence of the short-range part of the impurity potential is weak and the wave functions of the p-states vanish in the direct vicinity of the impurity because of the symmetry. The agreement between the theory and experiment for the excited levels was found to be so good that Faulkner[113] was

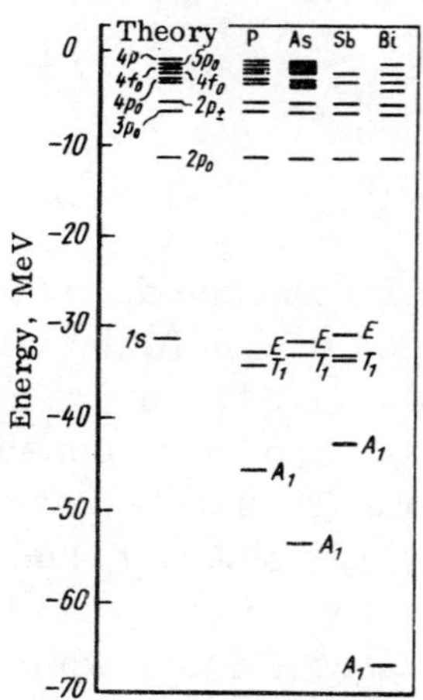

FIG. 51. Energy spectrum of donors in silicon.[113]

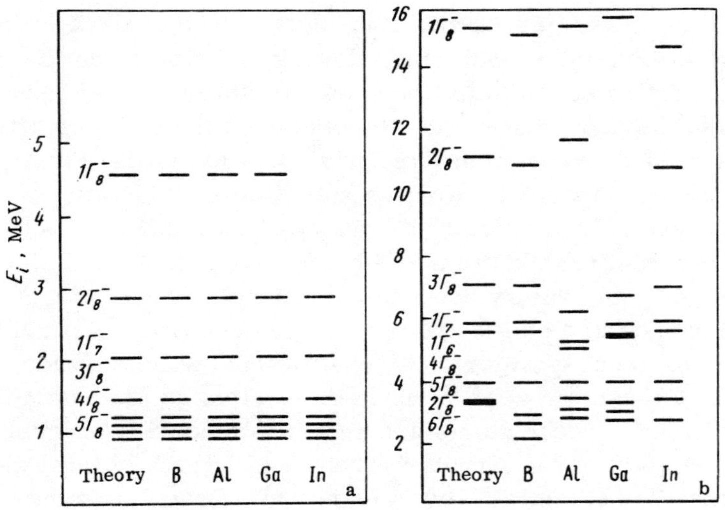

FIG. 52. Spectrum of excited states of acceptor impurities in germanium (a) and in silicon (b) (Ref. 101).

able even to refine the low-temperature values of the permittivity of germanium and silicon.

Equally good agreement between calculations and experiments has been achieved also for the excited states of acceptors in germanium and silicon. We shall not consider details of the theoretical model used in this case, but simply give the results[101] of calculations of the spectrum of the excited states of group III acceptors and the experimental results which convincingly demonstrate the good agreement (Fig. 52).

In the case of the ground-state energies of the impurity centres the agreement between the calculated and measured values is far from satisfactory. The discrepancies are frequently described by the "chemical shift" $\Delta = E_i - E_1$, where E_i are the experimental values and E_1 is the binding energy calculated from Eq. (23) assuming that $n = 1$. The chemical shift is associated primarily with the intervalley interaction and, naturally, with the special nature of the potential (electron structure) of an impurity centre. One of the more important corrections to the EMM relates to the polarization of a medium which near an impurity ion cannot be described by the macroscopic permittivity. In fact, an allowance for the spatial dispersion $\varkappa(r)$ alters radically the effective potential of an impurity in the central cell.

Among the numerous attempts to develop the EMM by refining the fundamental model and allowing for various corrections, we shall mention specifically the work of Pantelides and Sah[114] who solved the Schrödinger equation and used the true perturbation potential V(r) deduced from first principles for shallow and, which is particularly important, for deep substitutional donors; in this way they were able to rethink the theory of the EMM.

In contrast to the traditional point of view that the EMM is applicable only to shallow hydrogen-like levels, Pantelides and Sah[114] demonstrated that it applies also to deep centres provided the impurity in question forms a substitutional solid solution and belongs to the same row of the periodic system as the host atom being replaced. These atoms are known as isocoric, because they have the same electron structure of the ion cores. In this case the perturbation potential V(r) does not have the short-range part. The similarity of the radial distribution $\psi(r)$ for the silicon ions and the isocoric phosphorus and sulphur impurities (Fig. 53) which create shallow and deep levels in silicon, enabled Pantelides and Sah[114] to calculate in the one-band approximation the energy spectrum of these centres without recourse to any fitting parameters. It is clear from Table XVII that the values of the energies of the ground and excited states of phosphorus and sulphur are in good agreement with the experimental results. For the other shallow

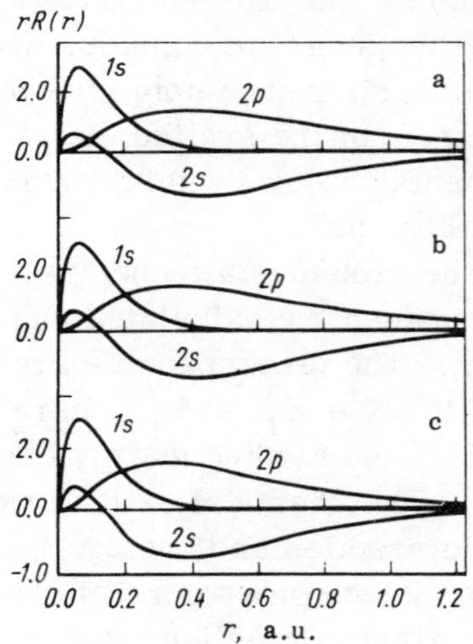

FIG. 53. Wave functions of ionic cores of silicon (a), phosphorus (b), and sulphur (c) (Ref. 114).

TABLE XVII. Ionization Energies of Donors in Silicon Calculated Using Generalized Effective Mass Method[114]

Impurity	r_l^*, a.u.	E_i, MeV	
		theory	experiment
Point charge	17.80	48.8	–
P	23.25	42.4	45.5
Singly charged two-charge centre .	5.46	1085.3	–
Neutral two-charge centre	6.43	489.0	–
S^+	6.80	659.3	613.6
S^0	7.60	297.1	302.0

* Localization radius.

donors of arsenic and antimony, which have ionic cores and R(r) very different from the cores and radial distributions $\psi(\mathbf{r})$ of silicon, the EMM is strictly speaking invalid[114] because the potential of nonisocoric (both shallow and deep) centres is large in the impurity cell (see, for example, Fig. 54) and does not satisfy the new condition of the range of validity of the EMM. In the generalized variant of the EMM the condition of smoothness of a potential of the (22) type is replaced by the requirement of the absence of short-wavelength components in the Fourier transform of the impurity potential. In this case the perturbation caused by the potential is small and does not mix the Bloch functions from different energy bands, thus justifying the validity of the one-band approach.

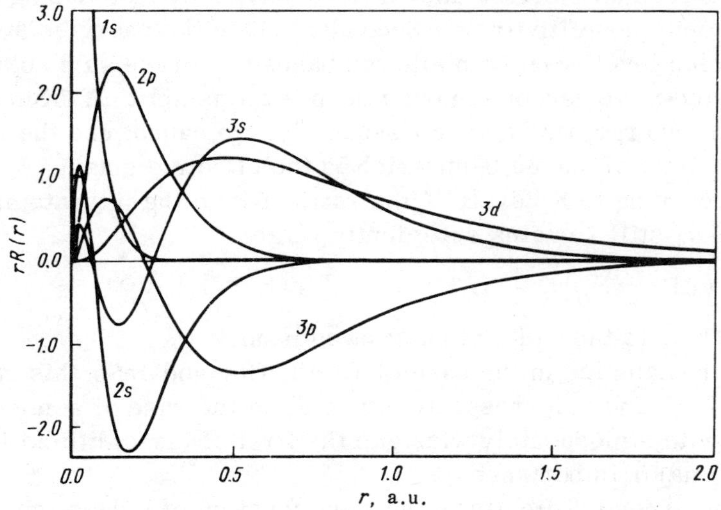

FIG. 54. Wave functions of the core of arsenic.[114]

We can therefore regard the work of Pantelides and Sah[114] as a successful attempt to rehabilitate the EMM and to extend it to the deep states of those substitutional impurities which have the same ionic core as the atoms they replace.

However, it should be pointed out that the approach used for the Ge:As system which is very similar to doped silicon fails to give such a good agreement with the experimental data.[114] This is explained in Ref. 115 by the lack of allowance for the umklapp scattering processes in the existing theories (including that of Pantelides and Sah[114]) because such processes play the dominant role in the intervalley mixing and govern the magnitude of the chemical shift. An allowance for these processes in the wave equation for a many-valley semiconductor makes it possible to achieve a better agreement between the experimental and calculated values of the energies of the split levels of the ground state of isocoric impurities in germanium and silicon.

A completely different approach to the problem of deep centres was developed in the EMM approximation by Keldysh.[116] In a description of deep states with a binding energy E_i of the order of the band gap E_G a theoretical treatment meets with two fundamental difficulties. Firstly, the electron orbit radius of a of the ground state of a deep impurity is much smaller than the radius of a shallow hydrogen-like centre. Therefore, it is doubtful whether it is possible to apply the EMM and to describe the interaction between an electron and an impurity ion by introducing a macroscopic permittivity $\varkappa$. Secondly, since the energies separating a deep level from both allowed bands are of the same order of magnitude, we cannot regard a deep level as split off from one particular energy band and, consequently, we cannot use the main characteristic of such a band which is the effective mass.

According to Keldysh,[116] the ratio of a to the interatomic distance a_0 still remains sufficiently large:

$$a/a_0 \propto (\Delta E_{all}/E_G)^{\frac{1}{2}} > 1,$$

where ΔE_{all} is the width of an allowed band.

For example, in the case of GaAs, Ge, and InSb this ratio is 2.6, 2.8, and 7.5, respectively, i.e. in the case of semiconductors with a moderately wide gap the first of these difficulties is not of major importance.

The second difficulty is the wave function of a deep impurity centre which should be a superposition of the wave functions of the conduction and valence bands, in the same way as the Bloch electron functions in the Kane model[117] are superpositions of the s–p states of the valence electrons of the host atoms. Keldysh[116] considered a solution to this problem. It should be stressed that in the derivation of the final formulae the problem was simplified by Keldysh by regarding the potential energy of the interaction of electrons with an impurity centre as purely of the Coulomb type.

The expression obtained for the energy of an impurity electron subject to the influence of both allowed bands characterized by a "quasi-relativistic" dispersion law $E(p) = \pm(\delta^2 + \delta p^2/m)^{\frac{1}{2}}$ is of the form*:

$$E_{n\gamma} = \delta/\sqrt{1 + \alpha^2/(n + \gamma)^2}, \qquad (26)$$

where $\delta = E_G/2$; $\alpha^2 = me^4z^2/\delta\hbar^2\varkappa^2$; $\gamma^2 = (j + \frac{1}{2})^2 - \alpha^2$; j is the total momentum; $n = 0, 1, 2, \ldots$.

In the ground state we have $n = 0$, $j = \frac{1}{2}$, and

$$E_0 = \delta(1 - me^4z^2/\delta\varkappa^2\hbar^2)^{\frac{1}{2}}. \qquad (26)$$

A comparison of this expression with the ground-state energy of an impurity obtained in the one-band approximation $E = \delta - \frac{1}{2}(me^4z^2/\varkappa^2\hbar^2)$ shows that inclusion of the second energy band results in significant "deepening" of the impurity level. The energies obtained from the two models discussed above agree only if $me^4z^2/\varkappa^2\hbar^2 \ll \delta$.

On the whole, the two-band model developed by Keldysh provided a qualitative explanation of the ability of a deep centre to capture carriers of both signs as well as some other characteristics of deep levels. However, the problem of the individual nature of the impurities (i.e. the dependence of E_i on the position in the periodic table of elements) is still unsolved because the impurity potential employed by Keldysh[116] is assumed to be the Coulomb potential $V = -ze^2/\varkappa r$.

The results of Keldysh[116] were used in quantitative calculations.[118–120] The ionization energy of acceptor levels in InSb was determined in Ref. 118 allowing not only for the conduction band, but also for all the valence bands (light- and heavy-hole bands) and for the band split off by the spin–orbit interaction. The value $E_i = 8$ MeV obtained by numerical solution was in satisfactory agreement with the ionization energy of shallow acceptors in this material. The same approach was applied somewhat earlier to

*The energies are measured from the middle of the band gap. When the spectrum E(p) is described in this way, the problem becomes formally similar to the familiar problem of an electron in a Coulomb field with relativistic corrections.

deep multiply charged acceptor centres in InSb and GaAs (Refs. 119 and 120), but the chemical nature of the impurity was ignored in both these investigations.

The two-band approximation with an allowance for the polarization correction to the impurity potential was used to calculate the energy spectrum of a deep acceptor level in InSb (Ref. 118). The experimental value of the ground-state energy of such an acceptor level was used in Ref. 116 to calculate the spectrum of excited states and the corresponding wave functions. It was also found that the wave function of a deep-donor electron is governed only by functions of the conduction band and ψ of a deep acceptor is determined by the wave functions of the conduction band and of the heavy-hole band. A similar result was obtained earlier by Tolpygo.[121]

The approach to the problem of deep acceptor states developed in Refs. 122 and 123 is similar to the two-band approximation. Ryabokon' and Svidzinskiĭ[122,123] allowed for the finite probability of penetration of a hole from the valence band into the inner (donor) shell of an impurity ion. Consequently, they found the wave function in the form of a superposition of two states representing a hole in the valence band and in the inner shell of the impurity. The system of equations can be reduced to a one-particle wave equation which contains not only the screened Coulomb potential, but also an effective short-range potential representing the chemical nature of the impurity. A comparison of the calculations with the experimental results for group III impurities in germanium shows that this approach is fully satisfactory. In the case of deep impurities Ryabokon' and Svidzinskiĭ[123] simply compared the order in the distribution of levels of transition elements in Ge and GaAs. However, they used the then available (and subsequently proved incorrect) experimental data for GaAs so that the agreement achieved between theory and experiment has become meaningless.

The two-band approximation can be regarded as an important refinement of the theory of deep local states, but we must bear in mind that the need to allow for the second energy band is not always justified. For example, an analysis of the case of deep donors in silicon (Fig. 55), Pantelides and Sah[114] drew attention to the fact that the contribution of the states at the top of the valence band at $k = 0$ to the wave function of deep donors is small for reasons of symmetry and the valence band states near $k = k_0$ lie relatively far from the impurity level. Since the energy band structure of

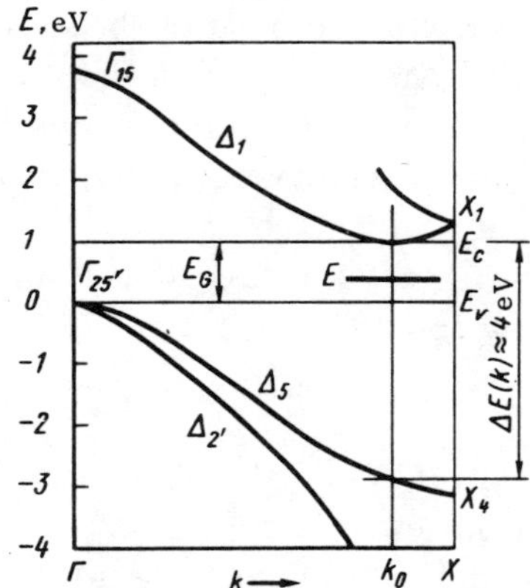

FIG. 55. Deep donor level of sulphur in silicon.[114]

silicon is of the indirect-gap type, the energy level of a sulphur impurity lies quite far from the valence band edge (only one of the six equivalent minima of the conduction band is shown).

Specific calculations indicate that in this case the influence of the valence band can be ignored. On the other hand, in the case of direct-gap materials the two-band approximation is essential in calculations of the energy spectrum of deep centres.

2.2. PSEUDOPOTENTIAL AND MODEL POTENTIAL METHODS

The work carried out using the pseudopotential method is of special importance in the theory of local centres. The basic idea of the method is a selection of the pseudopotential which is a smoother function of the coordinates and has a lower amplitude than the true potential, since this is particularly important for the solution of the problem under consideration within the perturbation theory framework. The wave function of an electron ψ can be represented in the form of a sum of a pseudopotential smooth function Φ describing the states between the cores of the atom and an oscillatory term ψ_s corresponding to states near the atomic cores:

$$\psi = \Phi - \sum_s \psi_s \int \psi_s \Phi d\tau.$$

It is appropriate to mention here that initially the pseudopotential method has been used to calculate the energy band structure of semiconductors. Since the smoothness of the potential acting on an electron is disturbed only in the direct vicinity of the sites in an ideal crystal lattice, the wave function of a free electron can be represented by a set of plane waves, whereas in the vicinity of the sites it can be represented by an oscillatory function similar to the wave function of a free atom.

The smooth part of the wave function of an electron is the eigenfunction of the equation

$$[(-\hbar^2/2m)\Delta + V_p]\Phi = E\Phi, \tag{27}$$

in which the pseudopotential $V_p = V + V_R$ represents a sum of the atomic potential V and the local potential V_R, the latter acting as a repulsive potential which influences strongly the pseudopotential V_p near the nuclei and makes it smooth throughout the crystal.

It is important to note that the eigenvalues of Eq. (27) are identical with the eigenvalues of the initial wave equation, and the eigenfunctions become smoother. Expansion of the band states of an electron in terms of plane waves is the reason why even a small number of components in a Fourier series provides a satisfactory approximation for the pseudowave function Φ.

It is found that a rigorous construction of the pseudopotential is a very difficult task. The pseudopotential method is used most frequently in the semi-empirical variant,[124] and in this case the fitting parameters of the theory are the Fourier components of the pseudopotential $V_p(\vec{r})$.

The paper of Phillips and Kleinman[124] was followed by widespread use of the pseudopotential method, which has become one of the most effective techniques for the calculation of the energy band structures of various semiconductors and metals. Ten years later the pseudopotential concept was used by Glodeanu[125] to calculate the energy spectrum of strongly localized impurity states in the framework of a two-band model. The potential of the interaction of a valence electron with an impurity centre was written down in the form of a pseudopotential V_p representing a sum of an impurity potential V and a local pseudopotential V_R: $V_p = V + V_R$. The wave function of an electron ψ in the conduction and valence bands should be orthogonal to the wave functions of the inner shells of an impurity atom. However, the final equations are very cumbersome. Glodeanu[125] suggested a different equation obtained in the one-band approximation and valid in the case of shallow and deep levels on condition that the impurity pseudopotential is correctly defined. This equation is similar to Eq. (21), except that the effective mass m and the permittivity $\varkappa$ depend on the coordinates. The function m(r) then has the following properties:

1) $m(r) \to m_0$ in the limit $V_p \to \infty$ (very deep levels);

2) $m(r) \to m$ in the limit $V_p \to 0$ (shallow levels).

In the intermediate case it is possible to find m(r) if the real pseudopotential is known. Consequently, we can write

1) $V_p \to V = e^2/\varkappa r$ and $\varkappa(r) \to \varkappa_0$ in the limit $r \to \infty$;

2) $V_p \to e^2/r$ and $\varkappa \to 1$ in the limit $r \to 0$.

Glodeanu reported in Ref. 126 specific calculations of the energy spectra of deep impurity states of copper, iron, cobalt, manganese, and nickel in silicon. A comparison of the results obtained with the experiments showed discrepancies, but Glodeanu pointed out that the occurrence of the calculated levels within the band gap is a success of the theory because of the roughness of the approximations which were made. However, the selection of the transition metal impurities for checking the pseudopotential theories clearly are inappropriate, because according to the results of Ref. 126, the electron structure of the d-levels cannot be described in terms of the conventional pseudopotential approach. In fact, the core of a transition metal is not sufficiently "rigid", in contrast to the centres with weakly bound s or p valence electrons and a core of completely filled shells. This means that the theory cannot be limited to the orthogonalization of the ψ function of a d-electron of an impurity to the wave functions of the band electrons, but a more complex calculation procedure must be used as was done – for example – in Ref. 127.

The pseudopotential method was developed further by Pantelides and Sah[128] who used the pseudopotential formalism and the many-valley approximation of the EMM to calculate the energy spectrum of various singly and doubly charged substitutional and interstitial impurities in silicon. The impurity pseudopotential was derived from first principles using the fundamental properties of a silicon crystal and the atomic properties of an impurity without any fitting parameters. Figure 56 shows, by way of example, the characteristics of the pseudopotential of a nonisocoric impurity (arsenic) in silicon, namely the effective charge $Z_{pb}(r)$ and its Fourier transform $\rho_{pb}(q)$, which are related to the pseudopotential of the impurity core V_{pb} by the expressions $V_{pb} = Z_{pb}(r)/r$ and $V_{pb}(q) = 4\pi e^2 \rho_{pb}(q)/q^2$, where q is the wave vector. For comparison, the dashed lines in Fig. 56 represent the same parameters for the Coulomb potential of a point charge. We can see that the pseudopotential is a much smoother function in the normal and reciprocal space compared with the true impurity potential.

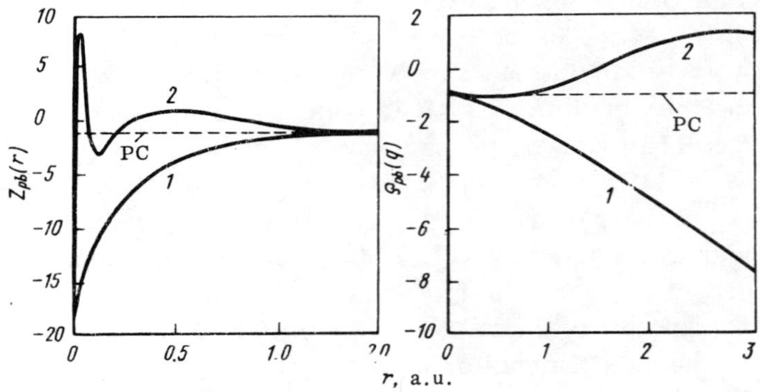

FIG. 56. "True" potential (1) and pseudopotential (2) of an impurity in Si:As (Ref. 101). Here, PC is a point charge.

The resultant wave function of an impurity electron was obtained in a many-band approximation and had the correct site structure in all the atomic cells, including the one containing an impurity. In the special case of isocoric impurities the theory gives an expression obtained by the same authors using a modified form of the effective mass theory.[114]

The results of a calculation of the energy spectrum of nonisocoric impurities in silicon and the experimental values of E_i are listed in Table XVIII. We can clearly see a satisfactory agreewith the theory and experiment, but the final judgement on the

TABLE XVIII. Binding Energies of Nonisocoric Impurities in Silicon Obtained by Pseudopotential (PP) and Model Potential (MP) Methods[128]

Impurity	E, MeV, obtained from		
	PP	MP	experimentally
		at interstitial positions	
Li	33.8	33.1	31.0
Na	34.6	32.4	–
Be^+	385.6	239.6	–
Be^0	146.5	103.6	–
Mg^+	259.0	190.2	256.5
Mg^0	98.0	81.3	107.5
		at substitutional positions	
N^0	335.9	27.0	–
N^-	52.5	0	45.5
As	53.1	40.7	53.7
Sb	31.7	35.3	42.7
Se^+	921.3	559.0	–
Se^0	358.4	265.3	–
Te^+	246.0	256.2	–
Te^0	71.9	117.7	–

quantitative correctness of the theory cannot be made without more experimental data.

The pseudopotential method is a valuable calculation technique if complete information is available not only on the electron states of the impurity ions, but also on the states of all the bands in an ideal lattice. However, such information is available only in a few cases. Moreover, as already pointed out, the pseudopotential method is in principle unsuitable for impurity ions with a partly filled shell.

A model pseudopotential method, developed by Abarenkov and Heine[129] and representing essentially a modification of the pseudopotential method, has recently become very popular. The main idea behind this method is to borrow potentials and wave functions of electrons at the inner shells of free atoms and ions on the assumption that they are not greatly affected in a crystal compared with the free state. It is important to ensure that the selected model potential allows automatically for the interaction effects and for the correlation between the valence electrons and those in the ionic core. According to Anderson,[130] the model potential of a free atom or ion acting on a valence electron is

$$V_M = -\sum_l A_l(E)P_l \quad \text{for} \quad r < R_M;$$
$$V_M = -Ze^2/r \quad \text{for} \quad r > R_M, \tag{28}$$

where A_l is a coefficient dependent on the energy and momentum (l is the orbital quantum number); P_l is the projection operator governed by the angular part of the electron wave function $Y_{lm}(\vartheta, \varphi)$; R_M is a free parameter representing a distance of the order of the ionic radius. For each l the model potential of Eq. (28) represents a rectangular well with a Coulomb "tail" (Fig. 57).

In the case of an impurity atom in a lattice (for example, arsenic in germanium) the model potential can be written in the form

$$V_M = \begin{cases} -\sum_l \Delta A_l P_l & \text{for} \quad r < R_M, \\ -Ze^2/r & \text{for} \quad r > R_M, \end{cases} \tag{29}$$

where $\Delta A_l = A_{l_{As}} - A_{l_{Ge}}$.

The subsequent procedure of solution of the problem reduces to plotting the screened potential of an impurity centre using Eq. (29) and allowing for the spatial dispersion of the permittivity so as to achieve matching of two solutions of the Schrödinger equation for

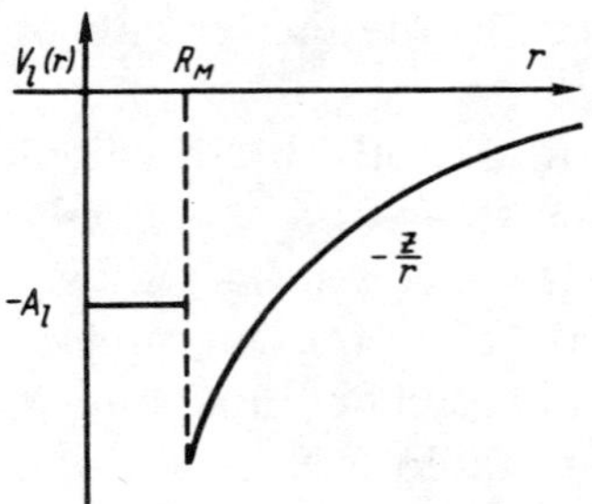

FIG. 57. Heine–Abarenkov model potential for atoms and ions.

$r < R_M$ and $r > R_M$ at the point $r = R$.

Specific calculations of the ionization energies of impurities carried out using the Heine–Abarenkov model potential in combination with the effective mass method were carried out[131] in the case of shallow group V donors in silicon. The agreement with the experimental results was achieved by introducing a certain dependence of the effective mass of an electron on the distance $m(\mathbf{r})$, which was earlier equal to the mass of a free electron m_0 for $r = 0$ and which now tends asymptotically to the known value m at large distances. The energies of the ground 1s state of phosphorus, arsenic, and antimony correctly reflect the tendency of changes found in the experimental values of E_i for these impurities, but there are serious doubts[101] about the approximations, particularly about the dependence $m(\mathbf{r})$.

The validity of the model pseudopotential method in the case of shallow and deep donors in silicon was considered in Ref. 108. It was found that the model potentials of doubly charged group VI donors (sulphur, selenium, and tellurium) are twice as large as the model potentials of the corresponding group V impurities (phosphorus, arsenic, antimony). The values E_i obtained by the pseudopotential and model pseudopotential methods for a large number of donors in silicon (Table XVIII) show that in the majority of cases the model pseudopotential method underestimates the relevant quantities quite considerably compared with the pseudopotential method and the experimental results. The calculated values of the binding energy of acceptors in germanium and silicon given in Ref. 132 also differ greatly from the experimental energies. Nevertheless, it is easier to construct a model potential than a pseudopotential and the use of a suitable model potential in combination with the effective mass method can become, according to Pantelides,[101] the most fruitful technique for the calculation of the properties of the majority of the impurity centres with the exception of elements in the first row of the periodic table, transition metal impurities, and lattice defects.

An original approach to the problem of deep states created by transition and noble metal impurities in semiconductors was developed by Fleurov and Kikoin[127] and independently by Haldane and Anderson.[133] The idea underlying this theory was put forward by Anderson[130] in order to describe the magnetic properties of systems comprising impurity atoms with a partly filled d-shell dissolved in simple metals.

The initial assumption made by Fleurov and Kikoin[127] is that in the case of a semiconductor at least one of the energy levels (for example, that of the n-th impurity electron) may be superimposed on the valence or conduction band of a crystal. An energy resonance between such a localized impurity state and the Bloch states of the continuum results, when an allowance is made for the s–d interaction, in hybridization of the impurity and host states. The wave function of an impurity centre is then described by

$$\psi(\vec{r}) = A[\psi_d(r) + \sum_{kn}[(d|Z|kn)/(E_i - E_{kn})]\psi_{kn}(\vec{r})],$$

where $\psi_d(\vec{r})$ is the wave function of an electron in the d-shell of an impurity atom; $\psi_{kn}(r)$ is a Bloch function; Z is the crystal field potential; E_{kn} is the energy spectrum of electrons in a crystal; E_i is the energy of a deep level.

The first term of the above relationship represents the localized part of a wave function and the second corresponds to a modification of the wave function because of hybridization. The characteristic localization radius $\psi_d(\vec{r})$ is of the order of the atomic radius, and the second term represents localization at much larger distances.

In contrast to metals, in which such an interaction broadens the d-level, but allowing for the correlation and formation of a localized magnetic moment, we find that renormalization of a d-level in a semiconductor by the hybridizing interaction has the effect that some of the d-electron states may be driven into the band gap (Fig. 58). Therefore, in accordance with this model, the resonance scattering of carriers by the d-levels of the impurities and an allowance for the collective effects are the main reasons for the appearance of deep localized states in the band gap of a semiconductor.

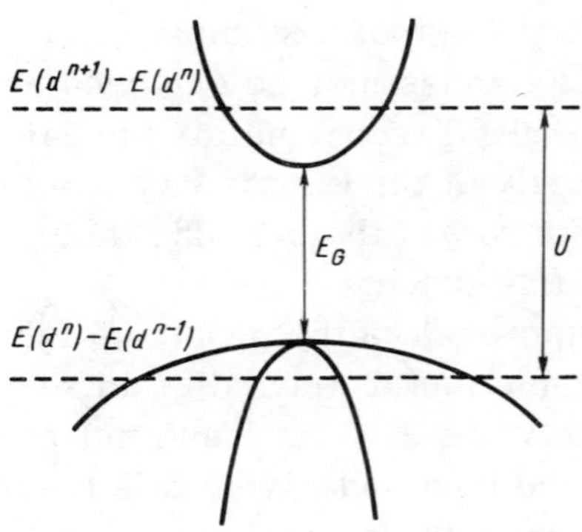

FIG. 58. Energy levels of the (n + 1)-th and n-th electrons corresponding to the difference between the terms $E(d^{n+1}) - E(d^n)$ and $E(d^n) - E(d^{n-1})$, respectively.[127] It is assumed that $U \gg E_G$.

It is worth mentioning that metals containing transition metal impurities cannot give rise to localized electron states because of the absence of a band gap near the Fermi level.

Estimates of the binding energy for deep levels[124] give quite reasonable values.

A similar approach to the problem of the nature of deep levels was used earlier.[122,123] However, no allowance was made for the theoretically important specific symmetry of the d-states which distinguishes T-metals from impurities with a filled d-shell.

However, we must bear in mind that generally we can encounter a situation when a renormalized level E_d does not fall within the band gap but remains inside one of the allowed bands. This may happen when a bare level E_d^0 is located much too far from the band edges because the matrix element of the hybridization with the states in the continuum is relatively small. It is understood that the theory fails to predict which specific case occurs in the investigated system. The answer can only be provided experimentally.

These ideas on the s–d interaction were developed further in Ref. 134. The authors allowed not only for the short-range potential, which mixes the states of a local centre and of the continuum, but also the Coulomb "tail" of the impurity core potential. This demonstrated that not only deep but also shallow s-levels can form. The resonance interaction then does not affect the energy position of the s-states in the band gap.

Special attention was given in Ref. 134 to an analysis of the symmetry of the impurity d-states allowing for the nonspherical part of the perturbation potential and for the real symmetry of a crystal. It was found that the resonance hybridization potential does indeed represent the covalent component of the crystal field of ligands and that deep states of the investigated impurities have the d-type symmetry and are classified in accordance with the irreducible representations of the point group of the crystal. The d-states split in the crystal field of the T_d symmetry and the proposed splitting scheme allows for the additional component of the ligand field associated with the hybridization. The results themselves can be regarded specifically as a rigorous justification that the semi-empirical model of the crystal field theory can be used to account for the observed optical and resonance properties of impurities with a partly filled d-shell in various crystals, including those characterized by a strong covalence.

It should be stressed that the treatment given in Ref. 134 represents the first attempt to develop a many-electron approach to the problem of deep centres. However, the results obtained cannot yet be compared quantitatively with experiments. Therefore, it is not yet clear whether the proposed model describes correctly the real situation and further checks are needed.

2.3. GREEN FUNCTION METHOD

Introduction of an impurity into a crystal has a consequence which is very important from the theoretical point of view: the potential energy loses its translational invariance. This greatly complicates the solution of the problem of the energy spectrum, but in the first approximation the point symmetry of an impurity centre in the lattice is not affected and all the electron states of a crystal with an impurity may be classified in accordance with the irreducible representations of the appropriate point group. An allowance for the local symmetry makes it possible to reduce greatly the order of the secular (time-dependent) equation which is, in the final analysis, obtained in the problem of determination of energy levels.

The adiabatic and one-electron approximations are insufficient for the development of a theory of an impurity centre in a crystal. Further significant simplification of the situation results from an expansion of the required wave function of the problem in a system of known functions, for example, the Bloch or Wannier functions of a perfect crystal, or atomic functions (atomic orbitals) of the environment. The last two forms of the basis of the expansion forming complete systems are of more local nature than the Bloch functions and they are therefore used particularly frequently in discussing deep-level centres within the framework of the Green function method and quantum-chemical approaches to the problem.

The many-body nature of the problem of an impurity centre in a crystal is only one of the difficulties. Another and clearly more troublesome is the selection of a suitable potential of a local centre. When this selection is made (for example, by the pseudopotential method), the problem reduces to the solution of the Schrödinger equation with a perturbation potential described by the operator $\alpha\overset{*}{V}$:

$$(H_0 + \alpha\overset{*}{V})\psi = E\psi, \tag{30}$$

where α is a parameter which is effectively an interaction constant.

The solution of this equation in the absence of a perturbation is known, so that the problem represents a search for a solution when $\alpha \neq 0$.

Further calculations of the energy spectrum are based on an expansion of the exact solution of Eq. (30) as a series in powers of the interaction constant and this can be done using the Green function method.

This method was applied specifically to the electron structure of impurity centres in crystals by Koster and Slater[135] and, strictly speaking, it represents a further development of the general theory of regular perturbations put forward by Lifshitz in 1947 (Ref. 136). According to this theory, the problem of finding the energy eigenvalues and wave functions of an impurity electron in a crystal reduces to a system of nonlinear homogeneous equations of infinite order. Fortunately, the rank of the corresponding secular equation can be reduced by a suitable selection of a system of basis functions that allow for the specific nature of the impurity centre. This is why Koster and Slater used the Wannier functions for impurities with deep levels in the band gap of a crystal.

The Green function method has a number of advantages compared with other methods for the calculation of the energy of local states. These advantages are basically as follows. Firstly, the Green function method makes it possible to calculate the energy spectrum of an impurity allowing for the many-body (many-particle) effects in all orders of perturbation theory. Secondly, it is possible to calculate equally accurately the average values of various physical quantities. For this reason the Green function method may, in principle, be used to treat different physical effects (electrical, optical, magnetic, etc.) with a high precision and from a unified standpoint.

Nevertheless, the Koster–Slater method is in practice used only in formal schemes. There are reasons for this. For example, a correct solution of the problem under consideration requires an allowance for a large number of terms in the expansion of the nearest sites in terms of the Wannier functions and the contribution of the wave functions of many energy bands to the required wave function of an impurity electron.

The problem finally reduces to the solution of a secular equation of the type

$$\det \| \delta_{nl} \delta_{R_\mu R_\nu} - V_{n\mu,l\nu} G_{n\mu,l\nu} \| = 0, \quad (31)$$

where $V_{n\mu,l\nu}$ are the matrix elements of the perturbation potential in the Wannier function basis; δ_{nl} and $\delta_{R_\mu R_\nu}$ are the Kronecker deltas; $G_{n\mu,l\nu}$ is the Green function.

The order of this equation is NN, where N is the product of the number of the investigated bands on the number of positions of the atoms in the environment of an impurity.

We can solve Eq. (31) if we have detailed information on the energy band structure of an ideal crystal which in the case of practically all the semiconductors is either unavailable or is very partial. Moreover, an increase in the number of sites in a crystal and of the points in the energy band structure which are included in the calculations increases greatly the rank of the corresponding determinantal equation so that a long time would be required for the solution even by the fastest computers. All this greatly complicates use of the Green function method in quantitative calculations relating to specific systems.

Nevertheless, Callaway and Hughes[137] determined the energy spectrum of a neutral vacancy in silicon by modelling the Wannier functions for eight energy bands and ten nearest sites, expressing these functions in terms of the Bloch functions and representing the potential of a defect by the pseudopotential of a silicon ion with the reversed sign: $V = -\lambda U_p(Si^{4+})$, where λ is a scaling factor.

The solutions obtained for the energy were found to depend strongly on the number of the included bands and on the value of λ; they were in full agreement with one another and with the experimental results.

Callaway and Hughes also considered a divacancy in silicon[138] and found that in this system there should be no bound states at all in the band gap, although this result contradicts the available experimental data.

A similar approach was also employed in Ref. 139 for an isoelectronic nitrogen impurity in GaP, but once again the agreement between the theory and experiment is not satisfactory.

In this connection we can only say that the limited success of the Koster–Slater method in specific calculations is probably not only due to the difficulties encountered in the use of the Wannier functions, but also due to insufficiently full allowance for the details of the band structure, as well as due to a number of other factors discussed earlier.

In addition to this calculation method, some progress has also been made in two variants of the Green function method. In the first of them an expansion of a Wannier wave function is carried out using the atomic orbitals of the valence electrons of a defect and of the atoms in the immediate environment: this is known as the Green function method in the linear combination of atomic orbitals (LCAO) representation. This approach to an allowance for the localization of deep impurity centres is closer to the situation under discussion than an expansion in terms of states of an ideal crystal, particularly in those cases when the orbitals of the impurity and host atoms are very different in respect of the symmetry (for example, impurities with a partly filled d- or f-shell in a crystal with the sp^3-hybridized bonds). Moreover, some advantages, compared with the conventional molecular orbital method, are imparted to the molecular orbital method in the LCAO representation by the formalism of the Green functions employed in this case. However, as in any other local approach, this method also suffers from difficulties. The main of these is associated with the selection of the optimal size of a cluster and of the corresponding boundary conditions. If we confine ourselves to just small clusters, such as those consisting of a defect and one coordination sphere, it is not possible to determine simultaneously the energy spectrum of an impurity crystal by "linking them" to one another. An increase in the size of a cluster results in increasing mathematical difficulties in the solution of the problem.

This variant of the Green function method was first used in specific calculations relating to a vacancy in diamond.[140] The basis was the same set of the s- and p-orbitals used to determine the energy spectra of ideal and defective crystals. The results obtained are definitely of interest although they are in poor quantitative agreement with the experimental results mainly because of clearly unreliable data on the energy band structure of the material. Two localized states of defects of the A_1 and T_2 symmetry have similar energies, although in reality the singlet state is in resonance with the valence band state and the triplet is in the band gap. It follows from calculations that the wave function of a defect is localized to the extent of 70% in the first coordination sphere and this is in good agreement with the experimental data.

Similar calculations for vacancies in germanium, silicon, and gallium arsenide[141] give results in good agreement with the experiments. Moreover, one of the results reported in Ref. 141 is of special interest: both allowed bands influence equally the energy position of a deep level in the band gap of a crystal. This is in full agreement with the approach of Keldysh to this problem, as already mentioned earlier.

Another variant of the Green function method is the continued fraction method developed for the calculation of the spectrum of electron states in amorphous semiconductors and, therefore, independent of the translational symmetry of a crystal.[142] The method was subsequently developed further by Masterov et al. in a study of the energy spectrum and wave functions, and of a number of parameters of the spin Hamiltonian of transition metal impurities in GaAs and GaP (Refs. 143 and 144). The initial model of the problem in these cases is an impurity quasi-atom (the term introduced by Il'in and Masterov[143]), the electron states of which can be represented by the LCAO of the impurity itself and of the atoms in the environment. The one-electron Green function of a defect crystal can be represented by a continued fraction and an allowance can be made for the properties not only of the immediate environment of a centre but of the whole crystal. In other words, the Green function formalism used here makes it possible to overcome the limitations of the simple cluster model and to allow for an important specific feature of an unbounded crystal. Another distinguishing characteristic of this method is that it does not require a preliminary determination of the energy band structure of the original (ideal) crystal.

The positions of the edges of the allowed bands of the host crystal are calculated using the relationship between the Green function $G_{ii}(E)$ and the local density of states $n(E)$ introduced by Friedel[145] to discuss the local effects:

$$n(E) = -(1/\pi) Im G_{ii}(E). \tag{32}$$

The condition n(E) = 0 determines the allowed band edges. The function $G_{ii}(E)$ is a continued fraction:

$$G_{ii}(E) = \cfrac{1}{E - a_1 - \cfrac{b_1}{E - a_2 - \cfrac{b_2}{E - a_3 - \ldots}}} \tag{33}$$

where the coefficients a_i and b_i are defined in terms of the elements of the original matrix in the LCAO basis. According to Ref. 143, the quantities a_i and b_i converge quite rapidly if the operator in the wave equation of an ideal crystal is the Weaire–Thorpe model Hamiltonian; in this case the continued fraction is readily summed. The self-consistent parameters of the selected Hamiltonian are found using some experimental data on NMR.

In the case of those impurity states which are inside the band gap it is possible to allow for the many-electron nature of the problem of localized electrons at the d-levels of an impurity in a crystal. This is done using the Hartree–Fock approximation and the basis in the form of solutions obtained in the one-electron approach.

The calculated values of the band gap of GaAs and GaP, the energies of the ground and lowest of the split-off (excited) states, and the values of the splitting Δ (see Table XIV) of T-metal impurities in these crystals are not in very satisfactory agreement with experiments.

However, this model accounts quantitatively for some of the experimental results on ESR, particularly a strong localization of the wave function of the impurity states, the shift of the g-factor, and the width of the ESR lines.

This continued fraction method has been used also to deal with deep centres due to gallium and arsenic or phosphorus vacancies in GaAs and GaP. According to the calculations, a neutral vacancy has acceptor levels and the ionization energy is $E_i \approx 0.3$ eV for gallium arsenide and gallium phosphide. On the other hand, arsenic and phosphorus vacancies create deep donor levels in GaAs and GaP with the binding energies 0.25 and 0.2 eV, respectively.

On the whole the method of continued fractions is quite difficult to apply in practice, because determination of the Green function and of the local density of states in a defect crystal requires an allowance for the relatively large number of fractions, which greatly complicates the calculations. Moreover, a determination of the energy levels of an impurity for a finite density of states in the band gap of a crystal is also a complex task. Finally, the continued fraction method ignores the band states (with the exception of the positions of the band edges) and these states affect strongly the energy spectrum of deep levels. This circumstance makes the continued fraction method a more qualitative rather than a quantitative method for the calculation of the electron structure of the strong localization centres.

We shall conclude this section by mentioning another possibility of determination of the local states based on the Green function method and the use of equivalent orbitals as the basis functions. The calculated spectra of neutral and charged vacancies, and also of isocoric impurities (sulphur and phosphorus) in silicon are, as in the case of the Green function method in the LCAO basis, in satisfactory agreement with the experimental results.

2.4. QUANTUM-CHEMICAL METHODS

The quantum-chemical methods are usually held to include the crystal field theory and many variants of the molecular orbital method which have largely appeared in the course of development of the quantum theory of chemical binding, structure, and properties of coordination compounds.[5]

The main features of the crystal field theory were considered by us in Chap. 1. Therefore, we shall simply caution the reader that many important problems such as the superhyperfine structure of the ESR spectra and other effects due to the transfer of spin density and delocalization of the wave function of an impurity centre cannot, in principle, be solved using the crystal field theory. Finally, this theory does not make it possible to "link" the energy spectrum of an impurity to the band structure of a crystal and thus carry out one of the most important tasks of the theory.

The molecular orbital (MO) method is a natural generalization in development of the crystal field theory. It allows for the electron structure of all the atoms forming a complex. A complex is regarded as a quasi-molecule composed of the atomic cores of impurity and its environment; the valence electrons move in the field of these cores. Naturally, the motion of each of the electrons depends on the positions of all the nuclei (in the first approximation these are assumed to be immobile) and all the other electrons. The exact solution of the Schrödinger equation of such a many-electron many-centre problem is obviously impossible because of the enormous mathematical difficulties. The solution to this problem is partly to use the one-electron approach when it is assumed that each electron moves independently in an effective field created by all the other charges. In this case the electron properties of a complex are described by a set of one-electron states with the wave functions extending over the whole system and the corresponding atomic orbitals form the molecular orbitals.

In all the variants of the MO method a crystal is modelled by a single fragment or a cluster composed of an impurity atom or a defect and of the host atoms. The subsequent solution of the problem depends on the selection of the shape and size of the cluster, and on the boundary conditions on its surface. Clearly, when a cluster is small and consists of just the central impurity atom and its immediate environment, it is impossible to determine the energy spectrum of a local centre relative to the edges of the band gap because the very concept of the gap of forbidden states loses its usual meaning.

On the other hand, an increase in the size of a cluster (number of coordination spheres) rapidly increases the number of atoms which must be included in calculations and, consequently, the volume of calculations frequently becomes impossible to carry out. Nevertheless, quantum-chemical models considered in the MO method or framework occupy an important place in the theory of deep centres.

There are many varieties of the MO theory, but we shall consider only the main ones.

Model of a Defect Molecule

This method was developed by Coulson and Kearsley,[146] who were the first to attempt to apply a local approach to the problem of a crystal with a defect by considering a vacancy in diamond. The model of a defect molecule was used subsequently[147] to study the spectrum of states of a vacancy and of a host interstitial atom of carbon in diamond, and also by Watkins[148] in a study of the radiation defects in semiconductors by the ESR method in connection with an analysis of the influence of the occupancy of molecular terms on the Jahn–Teller effect.

This method is based on the concept of a small quasi-molecule consisting of central atom (defect) and of the atoms in the immediate environment. It is important to stress that only the bonds directed toward the defect are included in the calculations and the wave function of the defect is described by just four broken (and not all sixteen) sp^3 hybrid orbitals of the four nearest carbon atoms. It is therefore not necessary to allow for what are called the surface effects on the boundaries of a cluster, which simplifies greatly the calculation procedure although it reduces the quantitative accuracy of the results obtained.

In the model of a defect molecule the first step is to calculate one-electron properties of a defect by the method of the self-consistent Hartree–Fock–Roothaan field and this is followed by a determination of the terms for a many-electron quasi-molecule. The MOs found in this way are used to construct the Slater determinants with all possible values of the spin states of the many-electron system and the problem of the configurational interaction is solved allowing for the selection rules.

The rigour of the calculations with an explicit allowance for the many-electron interaction in the model of a defect molecule provides an opportunity for calculating the energy levels and energies of electron transitions, including the case when the Jahn–Teller relaxation of the lattice takes place around a defect,[149] and to explain quite satisfactorily the ESR spectra. However, the restrictive nature of the initial assumption of complete localization of the wave function of a defect at the nearest neighbours clearly results in an overestimate, compared with the true value, of the splitting of many-electron multiplets.

The model of a defect molecule ignores completely a perfect crystal. Therefore, many important problems relating to the determination of the local levels relative to the edges of the allowed bands, identification of their nature, and determination of the capture cross-sections and other properties of defects in a crystal are outside the scope of this model. These problems can be solved by consideration of both a perfect crystal and a crystal with a defect.

Molecular-Cluster Approximation based on MOs in the LCAO Representation

This approach to the problem of deep centres in a crystal was developed by Messmer and Watkins.[150] As in the model of a defect molecule, a crystal with a defect is represented by a fragment (cluster) but with a large number of coordination spheres.

In this approximation the wave function of a molecular orbital is represented by a linear combination of atomic orbitals:

$$\Psi = C_1\psi_1 + C_2\psi_2 + \dots + C_n\psi_n, \qquad (34)$$

where n is the number of atoms in a cluster.

The problem of finding one-electron molecular orbitals is reduced as usual to a solution of a secular equation:

$$\Sigma C_k(H_{ik} - ES_{ik}) = 0, \quad i = 1, 2, \dots, n, \qquad (35)$$

where $S_{ik} = \int \Psi_i^* \Psi_k d\tau$ is known as the overlap integral of the orbitals of atoms i and k; $H_{ik} = \int \Psi_i^* \hat{H} \Psi_k d\tau$ is called the resonance integral.

Equation (35) is of the n-th order in terms of the energy of the system. Its solutions are n different values of the energy E with a corresponding set of coefficients $C_1, C_2, \dots, C_n$ for each of them. In turn each such set of coefficients defines its own molecular orbital and the total number of these orbitals for the system is n.

The number of atoms in a cluster acceptable in the calculation of Eq. (35) is set by the capabilities of modern computers; in the case of complexes of the cubic symmetry with the s- and p-orbitals it is necessary to limit the dimensions of a cluster to six coordination spheres ($n \leq 71$).

It is natural to solve the problem using the framework of the self-consistent field method mentioned earlier, but in the case of small molecules this method becomes too cumbersome. It is therefore necessary to search for simpler calculation methods. One of them is known as the extended Hückel theory (EHT), which is used widely in dealing with coordination compounds. It is important to note that in the EHT approximation all the matrix elements in the wave equation are defined in terms of the potentials of atoms and the integrals of the overlap between the atomic functions.

The energy band structure of a crystal considered within the EHT framework is described by one-electron levels of a selected cluster forming two groups, which are attributed to the valence band (bonding MOs) and to the conduction band (antibonding MOs). The energy gap between these two groups of states is then regarded as the band gap of a crystal. However, we can see from Fig. 59 that the calculated value of the band gap and the energy level scheme corresponding to the conduction band depend strongly on the dimensions of a cluster.

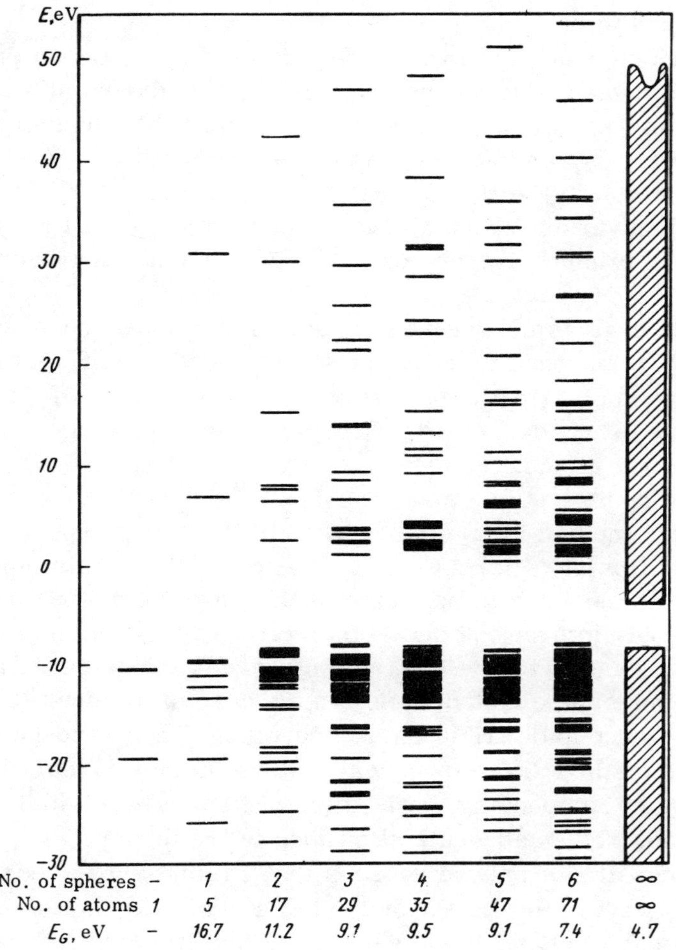

FIG. 59. Energy spectrum of a diamond cluster shown as a function of the number of the coordination spheres around the central carbon atom.[150] The calculated energy band structure of a diamond single crystal is shown on the right.

The results of many investigations show that the energy positions of impurity levels also vary considerably when the number of atoms in a cluster is increased and the dependence of E_i on the size of a cluster is nonmonotonic. For this reason alone the possibility of a correct determination of the energy spectrum of impurities within the framework of the cluster approach based on the EHT method is doubtful.

A serious problem in the models under consideration is a correct allowance for "broken" bonds at the boundaries of a cluster. This problem was solved by Messmer and Watkins[150] by saturation of the outer bonds of surface atoms in a cluster by additional electrons, but this disturbs the electrical neutrality of the whole system. Another variant which allows for the boundaries of a cluster involves saturation of broken bonds by hydrogen atoms ("hydrogen saturation").[151]

In this case an increase in the size of a cluster increases greatly the time needed to solve the secular equation of the problem. For example, "hydrogen saturation" of a cluster of 35 carbon atoms (we shall denote it by C_{35}) gives rise to a larger cluster consisting of 71 atoms ($C_{35}H_{36}$).

These complications make the EHT a qualitative rather than a quantitative method for the calculation of the energy spectrum of defects in a crystal. Moreover, in this approach the total energy of a system is assumed to be a sum of the energies of one-electron states, i.e. the interaction between electrons is assumed to be compensated exactly by the interaction between the nuclei in a cluster. Thus can be justified only in the case of an ideal cluster with a homogeneous distribution of electrons. The presence of a defect in a crystal results in a local distortion (relaxation) of the lattice and a corresponding distribution of the charge, which naturally requires an adequate allowance in the theory.

It is possibly for these reasons that attempts have been made[152–154] to apply the LCAO method to a cluster with a defect supplemented by the use of the self-consistent field procedure. This makes it possible to describe more correctly the distribution of the charge near a defect and the total energy of a system as a function of the positions of atoms in a cluster and also allowing for the electron–electron repulsion.

Nevertheless the specific results obtained for boron and nitrogen impurities in diamond[152] and for copper and silver impurities in some II-VI compounds[153,154] leave much to be desired.

Among the other cluster approximations for the calculation of the electron structure of defects in crystals, we shall mention also the method of scattered waves (SW-X_α),[155] which makes it possible to construct a unified energy spectrum of a defect in a crystal, which is particularly important in the theory of optical transitions and transport phenomena.

This method, which is also based on the self-consistent field

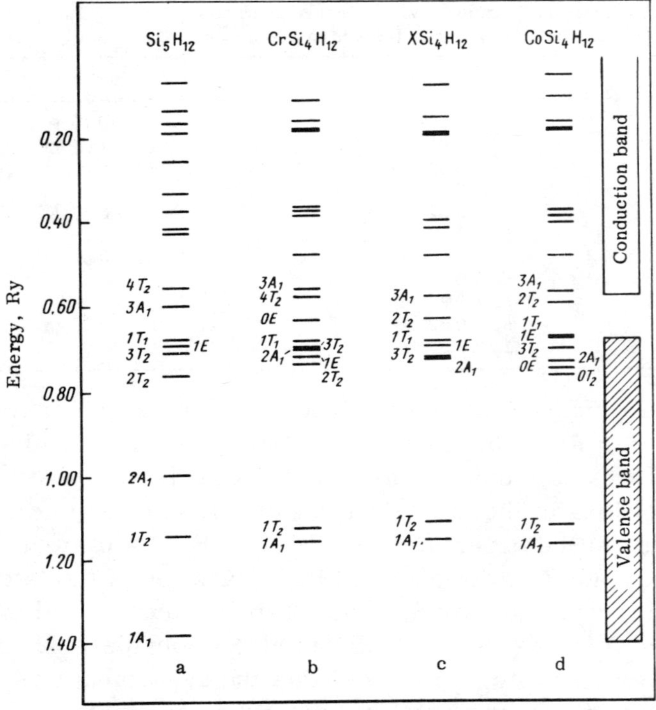

FIG. 60. Energy spectra of clusters corresponding to a perfect silicon crystal (a), to Si:Cr (b), to a vacancy in silicon (c), and to Si:Co (d).

procedure, was used[156] to calculate the spectra of the local states of ions with a partly filled 3d-shell present as impurities in silicon. A cluster of five silicon atoms (the central atom plus those in the first coordination sphere) was considered and it was assumed that the broken bonds are saturated with hydrogen atoms (Si_5H_{12}). A crystal with a defect was modelled by a cluster XSi_4H_{12}, where X is an impurity which alters the states in a perfect crystal (Fig. 60). According to the calculations, all the levels in the spectrum of a perfect cluster Si_5H_{12}, including the level $1T_1$ at the energy −0.676 Ry or lower, are completely filled and the levels $3A_1$ at the energy −0.598 Ry and higher are completely empty. The energy gap between the free and occupied states is 0.078 Ry or 1.05 eV, which unexpectedly agrees with the value of E_G = 1.15 eV for silicon at 0 K. The calculated values of the ionization energies of chromium, cobalt, nickel, copper, and zinc are listed in Table XIX together with the experimental results.

TABLE XIX. Ionization Energies (eV) and Nature of Levels Formed by Impurity Ions with Partly Filled d-Shells in Silicon[156]

Impurity	Experiment	Theory
Cr	$E_V + 0.74$ (D)	$E_V + 0.70$ (D)
Co	$E_V + 0.62$ (A)	$E_V + 1.03$ (A)
	$E_V + 0.52$ (A)	$E_V + 1.03$ (A)
	$E_V + 0.35$ (D)	–
Ni	$E_V + 0.82$ (A)	$E_V + 0.83$ (A)
	$E_V + 0.23$ (A)	–
Cu	$E_V + 0.52$ (A)	$E_V + 0.57$ (A)
	$E_V + 0.37$ (A)	–
	$E_V + 0.24$ (A)	–
Zn	$E_V + 0.60$ (A)	$E_V + 0.26$ (A)
	$E_V + 0.31$ (A)	–

The agreement between the theory and the experimental results obtained in Ref. 153 should be regarded as quite nominal because there was no indication as to which charge states of impurities and positions in the lattice (at regular sites or at interstices) corresponded to the experimental values of E_i. Moreover, the calculations made no allowance for the relaxation of the lattice around local centres or for the Jahn–Teller effect typical of ions with a partly filled d-shell, which can affect considerably the energy of impurity levels. In view of this the agreement was even more surprising. In any case, an attempt to calculate the spectrum of Cr^{2+} in GaAs carried out by the same authors[154,155] using a similar approach was clearly unsuccessful from the point of view of the agreement between the theory and experiment. The SW-X_α method was also used in Ref. 157 to calculate the electron structure of T-metal impurities in GaAs. The authors considered the properties of copper, nickel, cobalt, and iron in clusters consisting of 17 atoms. However, once again the basic theoretical result is the existence of deep levels formed by these impurities. Acceptable quantitative agreement between the calculated and experimental values of E_i was not achieved.

A general characteristic of all the discussed variants of a local description of a defect in the MO approximation is lack of allowance for the translation symmetry. This circumstance is responsible for many difficulties of the theory, particularly the already mentioned problem of the boundary conditions even in the case of large clusters. This fundamental problem can be solved much more successfully, as pointed out above, by combining a local approach with the Green function representation in the form of an expansion in terms of the Bloch states of an ideal crystal (LCAO method in combination with the Green function method).

2.5. SEMI-EMPIRICAL VARIANTS OF THE THEORY OF DEEP CENTRES

In addition to the calculation methods discussed above, various semi-empirical approaches are used widely in the theory of local states. The basic idea is to express certain characteristics of an impurity centre and its environment in a crystal in terms of other which are known from the experimental results.* Usually the fitting parameter, taken from the experimental data, is the ground-state energy E_i and the approximate wave functions of an impurity centre are selected so that they describe correctly the observed effects (for example, the absorption spectrum in the impurity region or the hyperfine interaction constants in ENDOR studies).

A semi-empirical approach of this kind is used in the well-known series of investigations of the spectral dependence of the photoionization cross-section of deep impurities carried out by Lucovsky.[60,159] Ignoring the "Coulomb correction" to the potential of a deep centre, the author describes it with the aid of the delta function: $V(\vec{r}) = A\delta(\vec{r})$, where A is governed by the ground-state energy E_i. The wave function of an impurity electron is then

$$\psi(r) = (1/2\pi a_1)^{\frac{1}{2}}(1/r)\exp(r/a_1), \tag{36}$$

where $a_1 = (\hbar^2/2mE_i)^{\frac{1}{2}}$.

In the dipole approximation of perturbation theory the dependence of the absorption cross-section on the photon energy for optical transitions from the valence band to a local centre is given by the expression

$$q_p(x) = 7\cdot 10^{-16}/[n(m/m_0)]f_1(x), \tag{37}$$

where $x = \hbar\omega/E_i$; $f_1(x) = (x-1)^{3/2}x^{-3}$; n is the refractive index.

In the case of transitions from local levels to the conduction band, the cross-section is

$$q_n(x) = 9\cdot 10^{-16}/[n(m/m_0)]f_2(x), \tag{38}$$

where $x = \hbar\omega/(E_G - E_i)$ and $f_2(x) = (x-1)^{\frac{1}{2}}x^{-1}$.

The nature of the dependences of the cross-sections defined by Eqs. (37) and (38) on the reduced photon energy $\hbar\omega/E_i$ can be judged on the basis of Fig. 61. The proposed theory describes

*According to Phillips,[158] theory is quite acceptable if the number of the parameters predicted by it is at least twice the number of the fitting parameters.

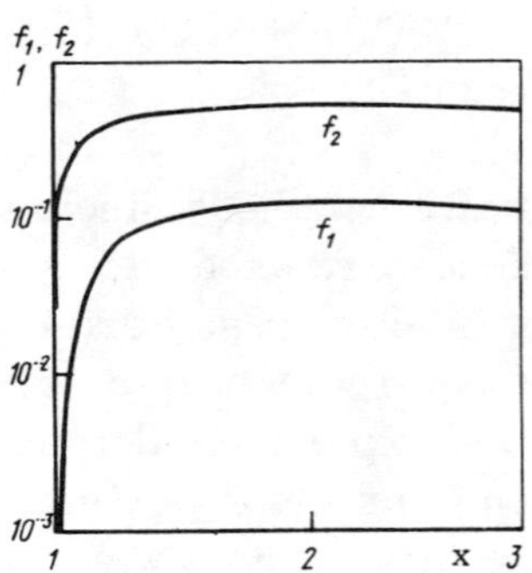

FIG. 61. Dependences of the functions f_1 and f_2 on the reduced energy.

satisfactorily the experimentally determined absorption spectra of various crystals containing deep-level impurities, for example, Si:In (Ref. 159), Si:Au (Ref. 160), GaP:O (Ref. 161), etc. However, there is no agreement at all between the theory and experiment for many other systems, including III-V compounds containing impurities of some transition metals.

Bebb[162] also reported theoretical calculation of the cross-section for impurity absorption within the framework of a more general model of a quantum defect. The wave function is taken to be the solution of the wave equation

$$[(-\hbar^2/2m)\Delta - (e^2/\varkappa r)]F(r) = E_i F(r), \tag{39}$$

valid when $r > r_0$, where r_0 is the range of action of the non-Coulomb part of the local potential of an impurity which depends on the nature of this impurity. In contrast to Eq. (21), in the case of the effective mass method it is assumed, as in the case of the Lucovsky theory, that E_i is the experimentally determined ionization energy of an impurity centre.

In the method of a quantum defect the wave function of an impurity electron can be represented in the form

$$F(r) = N_\nu r^{\nu-1} \exp(-r/\nu a), \tag{40}$$

where a is a Bohr radius and ν is a fitting parameter described by the expression $\nu^2 = -Ry^*/E_i$.

We can see that in the limit $\nu \to 1$ (shallow level case) the function $F(r)$ becomes similar to the analogous wave function considered in the effective mass approximation. On the other hand, in the limit $\nu \to 0$ (case of a deep level) the function $F(r)$ becomes similar to the solution of a problem with a δ-like potential.

Although in the range of small values of $\vec{r}$ the wave function of a defect remains indeterminate, the expression obtained for the photoionization cross-section in the method of a quantum defect is in good agreement with the experimental results, for example, those for acceptor impurities in the form of group III elements in silicon.[162]

The semi-empirical approach to the determination of the cross-section for the absorption of light by deep centres was developed further by Allen,[163] Jaros,[164] Langer, and others.[165,166]

Calculations of the frequency dependence of the absorption coefficient $q(\hbar\omega)$ allowing for hybridization of deep d-type impurity levels were published recently.[167] The hybridizing interaction between the local and band states results in the loss of parity by the impurity states and this has a considerable influence on the nature of the transitions and of the functions $q(\hbar\omega)$.

The expressions obtained for the optical transitions to both allowed bands from levels of the Γ_{12} symmetry are in fact close to the simple formulas (37) and (38); the differences are due to an allowance for the energy band nonparabolicity and for the hybridization-induced distortions in the wave function of an impurity electron. However, in the case of the states with the Γ_{15} symmetry the expressions for $q(\hbar\omega)$ are found to be much more complex, in spite of the fact that – for the sake of simplicity – no allowance has been made for the participation of phonons in optical transitions or for a considerable modification of the energy structure of the impurity levels because of the spin–orbit interaction and because of the Jahn–Teller effect. As pointed out in Ref. 167, an allowance for these factors should make it possible to develop a consistent theory of optical transitions from deep d-states.

Interesting results were obtained by Ridley[168] from a calculation of the impurity absorption spectra allowing for the contribution of the central cell. The model employed, known as the billiard-ball model, draws attention to the close link between the photoionization and scattering cross-sections of an impurity and allowance is made for the size and charge of a defect in calculating $q(\hbar\omega)$. Moreover, it is assumed that the wave function of a localized state can be determined using the Bloch functions of several extrema.

The impurity potential is modelled in the spherically symmetric form shown in Fig. 62. In the region of the core $0 < r \leq r_0$ the attractive potential is $V(r) = V_0$, whereas in the region of the

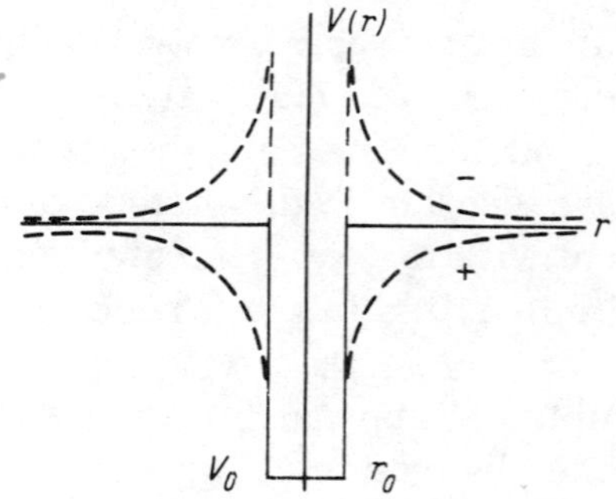

FIG. 62. Model potential of an impurity.[168]

Coulomb correction we have $V(r) = -Ze/\varkappa r$ for $r_0 \le r \le r_s$, where the effective mass approximation is valid, but $V(r) = 0$ for $r \ge r_s$ (r_s is the distance of the order of the screening length, not shown in Fig. 62).

It is assumed that the wave function of an impurity centre is of billiard-ball shape near the centre, so that $\psi(\mathbf{r})$ = const inside the core and $\psi(\mathbf{r}) = 0$ outside the core. It is assumed that $\psi(\mathbf{r}) = \Phi(\mathbf{r})S(\mathbf{r})$, where $\Phi(\mathbf{r})$ exhibits a strong dependence on the coordinates, compared with $\Phi(\mathbf{r})$, whereas $S(\mathbf{r})$ exhibits a weak coordinate dependence.

The problem of allowance for the central cell in $\psi(\mathbf{r})$ is solved making another assumption: depending on the origin of the level, the function $\Phi(\mathbf{r})$ is described by a Bloch function of one of the allowed (conduction or valence) bands or by a superposition of such functions.

The envelope function in the region of the core is described by a spherical Bessel function of zeroth order $S(r) = B_0[\sin(\alpha r)]/\alpha r$, where B_0 is the normalization constant and α is a characteristic fitting parameter dependent on the depth of the potential well V_0 of an impurity and on the energy E_i: $\alpha^2 = 2m(V_0 + E_i)/\hbar^2$. Outside the core the function $S(\mathbf{r})$ is described by

$$S(\mathbf{r}) = B_1(r/a_z^*)^{\mu-1}\exp[-r/\nu a_z^*],$$

where $\nu = [Z^2(Ry^*/E_i]^{\frac{1}{2}}$; $a_z = a/|z|$; $\mu = \nu Z/|Z|$.

The final expression for the photoionization cross-section is $q = q_0G(x)$, where $x = \hbar\omega/E_i$. In the case of a parabolic band near the absorption edge ($x \approx 1$) the spectral dependence $G(x)$ is of the form shown below:

Charge	$Z = 0$	$Z > 0$	$Z < 0$
Donor	$(x-1)^{3/2}/x$	$(x-1)/x$	$[(x-1)/x]\exp[-2\pi\gamma/(x-1)^{1/2}]$
Acceptor	$(x-1)^{1/2}/x$	$1/x$	$(1/x)\exp[-2\pi\nu/(x-1)^{1/2}]$

The conclusion that the calculation of $q(\hbar\omega)$ must allow for the influence of the core of an impurity is clearly one of the principal results of the theory. In fact, in other models, such as the model of a quantum defect, the behaviour of the wave function near $r = 0$ is of no importance; in this model the main role is played by the asymptote of $\psi(r)$ and it is this asymptote which in turn is ignored in the billiard-ball model.

As expected, Ridley found[168] that the two models complement each other to a large extent. However, the final judgement on the correctness of Ridley's theory can only be made by a comparison with experimental data. This is still to be done.

2.6. GENERAL STATUS OF THEORETICAL REPRESENTATIONS

All these methods for the calculation of the energy spectrum allow us to draw reliably just one conclusion that T-metal impurities form deep levels in the forbidden band gap of a semiconductor. The above account shows that at present there is no single theory of deep centres capable of describing sufficiently accurately all the variety of the properties of these centres in crystals. In this sense we agree with Roĭtsin[169] who answers the question "Does a theory of deep centres exist?" by pointing out that there seems to be a sufficient number of methods and the literature on the subject is impressive, "yet the trivial problem of the positions of energy levels, which is the touchstone of the theory of deep centres, has not yet been solved." It is therefore not surprising that the progress made in the description of other characteristics of centres of strong localizations is even more modest.

Among the existing methods the preference is clearly being given to semi-empirical approaches because more rigorous theories meet with major fundamental and calculation difficulties even in the simplest cases. Pantelides et al.[101,170] are of the opinion that the Koster–Slater determinant method in the LCAO representation is the most promising, although this conclusion is drawn more on the basis of the promise of such an advantage of the method than on the basis of the limited success already achieved in specific calculations.

The absence of a consistent quantitative theory does not mean at all that the efforts of the majority of the authors have been useless. Quite the opposite, they have made it possible to formulate the main ideas on the specific properties of deep local states in semiconductors and in any case to account qualitatively for the major part of the experimental observations. Moreover, the many difficulties facing the future theory have become clearer and this theory will clearly require the development of a microscopic many-electron approach allowing for the polarization, electron–electron interaction (including exchange), interaction with the environment and lattice vibrations, and other effects. A rigorous solution of such a very complex problem will undoubtedly meet with enormous computational difficulties and can hardly be used in specific calculations. The solution of the problem is probably to continue the search for more specific but relatively simple theories that allow for the main types of interaction in the system under consideration. This has not yet been done, but attempts to forecast the properties of deep centres in crystals have not always been successful and reliable values of their main parameters (energy spectrum, photoionization, capture, and scattering cross-sections of charge carriers, form in which impurities are found in the lattice, g-factor, solubility, distribution and diffusion parameters, etc.) can only be found experimentally.

Under these conditions it is very important to investigate experimentally local centres in sufficient simple (model) systems, the results of which can be compared in a more natural manner with the results of these or other theories. We must bear in mind that in selecting these model systems the degree of localization of an electron at a centre may vary within wide limits depending on a number of factors. These factors include: a) difference between the potentials of an impurity and the atom it replaces; b) screening effects $\varkappa(r)$ near a centre; c) lattice deformation because of the difference between the dimensions of an impurity and the host atom; d) influence of additional extrema of the energy bands, etc. In view of this the impurities in the form of transition metals in elemental semiconductors and in III-V compounds in general, and in GaAs in particular, are of considerable interest from the point of view of the theory. This is due to the following reasons.

1) The common feature of all the elements in the transition group is the partial occupancy of the d-shells. This is clearly the main reason for the strong localization of carriers at centres formed by these impurities.

2) In gallium arsenide, by analogy with elemental semiconductors germanium and silicon, the impurities in the form of chromium, manganese, iron, cobalt, or nickel create deep localized states in the band gap. However, in contrast to germanium and silicon, in GaAs these elements: a) are substitutional rather than interstitial impurities; b) exhibit a much higher (by 2-3 orders of magnitude) solubility; c) occur mainly only in two (+2 and +3) charge states (see Chap. 1).

3) Ions of impurities with a partly d-filled shell retain their individuality, typical of free ions, but they nevertheless are very sensitive to the immediate environment. Therefore, an impurity of the original transition metal acts a "probe" suitable for the investigation of very fine details of the semiconductor matrix itself by investigating spectroscopic effects in the rf and optical frequency ranges.

4) A fairly large number of transition-metal impurities makes it possible to study the chemical tendency in the positioning of the energy levels in the band gap and other properties of these metals; information of this kind is of considerable importance for the development of the main ideas on strong-localization centres in semiconductors. In fact, in view of the unavoidable simplifications in the original model, even a perfect theory cannot describe the energy spectrum of defects in a crystal with the required precision (0.1 eV). Therefore, the correctness of the theory is best determined not by a comparison of the calculated and experimental values of E_i for some randomly selected crystals, but by comparing a tendency to a change in energy levels of a group of impurities which to some extent combine the important properties. Transition-metal impurities with a partly filled d-shell are therefore a very suitable model system for local centres that create deep levels in the majority of semiconductors.

3. ENERGY SPECTRUM OF DEEP-LEVEL T-IMPURITY CENTRES. EXPERIMENTAL RESULTS

3.1. ENERGY LEVELS OF T-METAL IMPURITIES IN GERMANIUM

The energy levels introduced by T-metal impurities in germanium have been investigated thoroughly in the first half of the fifties.[171–174] These investigations were carried out so thoroughly that the reference data obtained at the time have not become obsolete. Only in some cases the subsequent studies have refined the values of E_t.

A summary of the ionization energies of the T-metal impurities in germanium can be found in Table XX. This table lists the

TABLE XX. Charge States and Ionization Energies of T-Metal Impurities in Germanium

Impurity atom	Charge state	$E_c - E_t$	$E_v + E_t$	Determination method	Reference
Cr	?	–	0.07	?	[175]
			0.12	–	–
Mn	A^-	–	0.16	PH; IQ	[171, 176]
	A^{2-}	0.37	–	R; PH;IQ	[171, 175]
Fe	A^-	–	0.34	R; PH; IQ	[171, 172, 175]
		–	0.25	?	[177]
	A^{2-}	0.27	–	R; PH; IQ	[172, 175]
		0.30	–	?	[177]
Co	A^-	–	0.25	R; PH; IQ; PHE	[174, 178]
	A^{2-}	0.31	–	R; PH; IQ; PHE	[173]
		0.30	–	?	[177]
	D^+	–	0.09	?	[177]
		–	0.081	?	[178]
		–	0.083	H; PH; GRN	[179]
Ni	A^-	–	0.23	R; PH; PHE	[174]
		–	0.20	?	[180]
		–	0.22	PZ	[181]
	A^{2-}	0.30	–	R; PH; PHE	[174]

Notes. 1. The following notation is used for the determination methods: R and H are the temperature dependences of the electrical resistivity and of the Hall coefficient, respectively; PH is the impurity photoconductivity; IQ is the infrared quenching of the intrinsic photoconductivity; PHE is the photo-Hall effect; GRN is the generation–recombination noise; PZ is the piezoresistance; D is a donor; A is an acceptor.

2. The levels were not determined in the case of scandium, vanadium, and titanium.

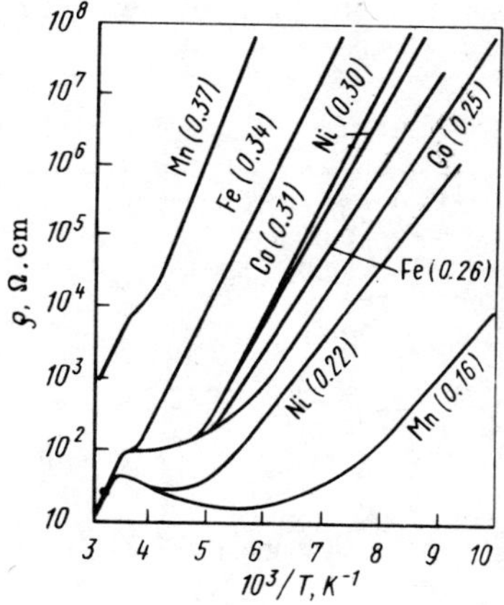

FIG. 63. Temperature dependences of the electrical resistivity of germanium samples doped with various T-metal impurities (slopes of the curves are given in parentheses).

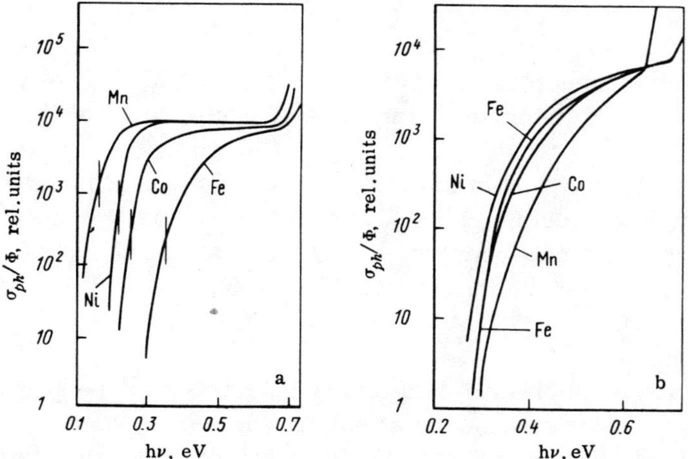

FIG. 64. Photoconductivity spectra of p-type (a) and n-type (b) germanium samples doped with T-metal impurities. A unit of the photoconductivity has the following relative values for the various elements:

Element	Co	Fe	Ni	Mn
Position a	10^{-6}	10^{-8}	10^{-6}	10^{-6}
Position b	10^{-8}	10^{-7}	10^{-7}	10^{-7}

charge states of a T-metal ion identified with one particular value of E_t. It is clear from Table XX that the majority of the Ge:T systems have been investigated on the basis of the temperature dependence of the equilibrium conductivity and of the impurity photo-

TABLE XXI. Charge States of T-Metal Ions and Their Electron- and Hole-Capture Cross-Sections in Germanium[182]

Charge state of impurity	$\sigma_n(T)$, cm^2	$\sigma_p(T)$, cm^2
Ni^0	10^{-16} (80–300 K)	–
Ni^-	$\sim e^{-0.02/kT}$ (80–100 K)	–
	$\sim e^{-0.01/kT}$ (100–200 K)	$T^{-1.5}$ (~80 K)
	10^{-19} (80 K)	10^{-13} (80 K)
	$6 \cdot 10^{-16}$ (300 K)	–
Ni^{2-}	–	10^{-14} (80–300 K)
Fe^0	10^{-15} (80–300 K)	–
Fe^-	$\sim e^{-0.05/kT}$ (80 K)	$\sim T^{-3}$
	$5 \cdot 10^{-19}$ (80 K)	$3 \cdot 10^{-15}$ (300 K)
Fe^{2-}	–	10^{-14} (300 K)
Mn^0	$2 \cdot 10^{-16}$ (80–300 K)	–
Mn^-	$4 \cdot 10^{-17}$ (100–300 K)	$3 \cdot 10^{-12}$ (60–80 K)
Mn^{2-}	–	$6 \cdot 10^{-17}$ (300 K)
	–	$\sim T^{-4.5}$
Co^0	10^{-15} (80–300 K)	–
Co^-	10^{-19} (80 K)	–
	10^{-17} (145 K)	–
	$3 \cdot 10^{-17}$ (300 K)	–
Co^{2-}	–	$\sim T^{-4}$
	–	10^{-14} (300 K)
	–	$2 \cdot 10^{-15}$ (145 K)

conductivity spectra. Therefore, Figs. 63 and 64 give these dependences for germanium samples containing some of the T-metal impurities.

Table XXI lists the electron and hole capture cross-sections of the various levels of transition-metal impurities in germanium.[182]

We shall now consider briefly the data on the various T-metal impurities in germanium by way of comments on Tables XX and XXI.

Scandium

There are no published data on any attempts to dope germanium with scandium. However, such attempts would have been desirable because it follows from Table IV that, irrespective of whether scandium atoms occupy substitutional or interstitial positions, they are trivalent in the nonionized state. In the former case they would have been similar to impurities from the third group of elements in the periodic system and would have formed shallow acceptor levels, whereas in the latter they would have then been of the acceptor nature, but would have introduced deeper levels into the band gap of germanium.

The crystallochemical approach (Fig. 6) shows however that the probability of formation of a substitutional solid solution in the Ge:Sc system is low.

Titanium

In contrast to scandium, there have been attempts to dope germanium with titanium,[175] but atoms of this element showed no electrical activity.

Vanadium

The study reported by Newman and Tyler[175] was also concerned with vanadium and, moreover, doping of germanium with vanadium by drawing ingots from the melt was reported by Woodbury and Tyler.[171] In a wide range of temperatures between room and 25 K there was no manifestation of electrical or photoelectric activity of vanadium impurity atoms. It should be pointed out that the double doping method was used by Woodbury and Tyler[171] to "drive" the Fermi level across the band gap of germanium and thus prepare n- and p-type vanadium-doped samples. The type of conduction was governed by the donor or acceptor nature of the hydrogen-like impurity introduced simultaneously with vanadium into germanium crystals.

An important result was the constancy of the temperature dependence of the mobility of carriers before and after doping of germanium with vanadium. This might be evidence of a very low concentration of vanadium atoms in the investigated samples. It is possible that the presence of vanadium in germanium is not manifested in electrical or photoelectric properties because of the low concentration. Unfortunately, labelled vanadium atoms were not used in the reported investigations. The use of such atoms would have made it possible to demonstrate the penetration of vanadium into the investigated samples during growth and to determine accurately the concentration of these atoms.

Chromium

Among all the T-metal impurities which contribute energy levels to the band gap of germanium least work has been done on chromium.

It is known only that chromium gives rise to two acceptor levels (Table XX) the energies of which are given in reviews[175–177] without any relevant details. Therefore, it is not clear which states of chromium in germanium should be identified with these levels. Moreover, it is not known whether it is possible for these levels to be present in equal concentrations.[175] Without such information we cannot tell whether the levels are correlated or independent.

Manganese

The energy levels of manganese were studied in germanium samples by the double doping method.[171] The crystals doped with manganese alone exhibited p-type conduction. The addition of arsenic introduced simultaneously with manganese produced samples with different positions of the Fermi level right up to a situation in which a conversion from p- to n-type conduction was obtained.

When the concentration ratio [As]:[Mn] was low, a shallow level at $E_v + 0.05$ eV was observed; according to the results of later investigations it could be identified, in our opinion, with an associated defect of the site arsenic–vacancy type. At some value of the ratio [AS]:[Mn], the low-lying levels, which were not related to manganese, were found to be completely compensated and the low-temperature resistivity of the samples increased by more than five orders of magnitude, which is clearly visible from Fig. 63. The slope of the $\rho(10^3/T)$ curve, amounting to 0.16 eV, and the p-type of conduction of these samples indicated that manganese formed an acceptor level with the ionization energy $E_v = 0.16$ eV. This level was not exhibited by control undoped samples.

When the concentration of arsenic was high, the samples exhibited n-type conduction and the dependence $\rho(T)$ shown in Fig. 63 had the slope $E_c - 0.37$ eV. This level was also attributed with a high probability to manganese.[171] This was supported particularly by investigations of samples with different manganese concentrations. Moreover, the same amounts of the arsenic impurity were required to compensate the 0.16 and 0.37 eV levels.

Impurity photoconductivity and quenching of the photoconductivity, as well as the photo-Hall effect were studied in the case of Ge:Mn:As samples.[171]

The photoconductivity spectra of p- and n-type Ge:Mn:As samples recorded in the impurity region are plotted in Figs. 64a and 64b. A good agreement was observed between the red edge of the photoconductivity and the two level energies E_t found from the temperature dependence of the electrical resistivity.

The sign of the photo-Hall effect was extremely important. It showed[175] that illumination of p-type samples with photons of energy $h\nu \gtrsim 0.16$ eV, i.e. photoionization of the lower level, released free holes. Photoionization of the upper level produced free electrons when n-type samples were illuminated with $h\nu \gtrsim 0.37$ eV photons. This result made it possible to identify both energy levels with, respectively, singly and doubly charged acceptor states of the manganese impurity in germanium (Table XX).

The reliability of this identification made it possible to explain in a nonconflicting manner the results of an investigation of the ESR spectra, as discussed in Sec. 1.3.

Iron

There have been no double doping investigations for the purpose of identifying the energy spectrum of iron levels in germanium. However, this approach was implemented accidentally[172] using iron supplied by two commercial firms. In one case the iron had a considerable amount of boron as an impurity and the other contained phosphorus. Therefore, it was possible to prepare p- and n-type samples, respectively.

The dependences $\rho(T)$ in Fig. 63 show that the n-type samples manifest a level at $E_c - 0.27$ eV, whereas p-type samples manifest a level at $E_v + 0.34$ eV. The error in the determination of the level positions was estimated by authors to be ±0.02 eV in both cases. The photoconductivity spectra of iron-doped samples with both types of conduction (Fig. 64) were fully analogous to the curves discussed earlier for Ge:Mn. Using the conversion coefficients given in the caption of Fig. 64, we can show that in the region of the plateau in the range $0.5 < h\nu < 0.7$ eV the inequality $\sigma^n_{ph} > \sigma^p_{ph}$ is obeyed. As shown in Ref. 25, the electron lifetimes in n-type samples were longer than the hole lifetimes in p-type crystals. Therefore, the photocurrents of the investigated samples obeyed the inequality: $j_n > j_p$. This inequality suggested that the $E_v + 0.34$ eV and $E_c - 0.27$ eV levels should be attributed to Fe^- and Fe^{2-}, respectively. In fact, in p-type samples the Fermi level is in the lower half of the band gap and the 0.27 eV level is always free of electrons. Illumination of such crystals with photons of energy $h\nu \approx 0.6$ eV transfers electrons from E_v to both impurity levels. This creates a current j_p which is governed by the number of vacant impurity levels. In the case of n-type samples the Fermi level is quite high in the band gap and practically all the impurity levels are filled with electrons so that the photocurrent is in this case higher than in the preceding case.

This identification of both levels with singly and doubly charged states of iron impurity ions is supported also by infrared quenching of the intrinsic photoconductivity observed on illumination by photons with the energy $h\nu = 0.37$ eV (Ref. 172) and also by the observation that the concentrations of centres contributing both levels were identical within the limits of the experimental error.[175]

Different energy values for iron were put forward later[177] without any explanation how they were determined (Table XX). These later values did not seem to be more reliable.

Cobalt

According to Refs. 173 and 175, cobalt forms two deep acceptor levels in germanium (Table XX) and these levels are attributed to the same centres in the singly and doubly charged states. The ionization energies of the two levels were found from the temperature dependence of the electrical resistivity in p- and n-type samples (Fig. 63). The different types of conduction observed on doping of germanium with cobalt were achieved either by single doping (p-type) or double doping (n-type), using in the latter case an Co + Sb alloy.

A study of the impurity photoconductivity (Fig. 64) showed that the edge energies E_t were in satisfactory agreement with the thermal ionization energies of each of the levels.

It was shown in Ref. 173 that n-type samples exhibited infrared quenching of the intrinsic photoconductivity on illumination with impurity-absorbed light. The quenching effect appeared at $h\nu$ = 0.32 eV, which was also in good agreement with E_t = 0.31 eV.

It should be pointed out that the infrared quenching effect was much weaker for Ge:Co than for Ge:Fe and, as demonstrated in Ref. 173, the relative change in the photocurrent as a result of the quenching effect was only 2% in the case of the cobalt-doped samples. Therefore, the quenching effect by itself was insufficient to regard cobalt as a doubly charged acceptor centre. However, this conclusion was supported by additional investigations of the photo-Hall effect[175] which demonstrated that in the case of n-type crystals the illumination with impurity-absorbed light released free electrons from the upper level, whereas in the case of p-type crystals such illumination released free holes because of the $E_v \rightarrow E_{t_1}$ transition.

Later[178] another level was observed and it was attributed to a donor state of cobalt in germanium. Its ionization energy was, according to different authors (Table XX), within the range 0.081-0.09 eV above the top of the valence band.

This level could not be attributed to a vacancy–donor associate, as was done in the case of Ge:Mn, because this was in conflict with the donor nature of the 0.081-0.09 eV level. Therefore, on the basis of the data in Fig. 25 it was possible to conclude only that the level corresponds to an interstitial $3d^8$ state of the ionized cobalt impurity in the germanium lattice and that both acceptor levels correspond to cobalt ions in substitutional positions.

Nickel

This is the most thoroughly investigated T-metal impurity in germanium. One of the first papers[174] reported a thorough study of the temperature dependence of the electrical resistivity of Ge:Ni:As and Ge:Ni:Sb with different ratios of the nickel and arsenic (or antimony) concentrations. Not only did the authors prepare samples with different types of conduction, but also p-type samples with different degrees of compensation, i.e. with different positions of the Fermi level.

This revealed two types of $\rho(T)$ curves as shown in Fig. 63 and made it possible to find two values of E_t shown in Table XX. The values of the ionization energy of the lower level differed only slightly from one author to another.

The impurity conductivity spectra of these samples are plotted in Fig. 64. In the case of the p-type samples with different compensations of the lower level it was possible to study the evolution of the photoconductivity spectrum as a function of the degree of occupancy K of this level (Fig. 65). At low values of K it was found that illumination with impurity light resulted in predominance of the filling of the $E_v + 0.23$ eV level with electrons from the valence band. The photoexcited carriers were holes, as confirmed by a study of the photo-Hall effect.[175] An increase in the degree of occupancy K resulted in double optical transitions that released

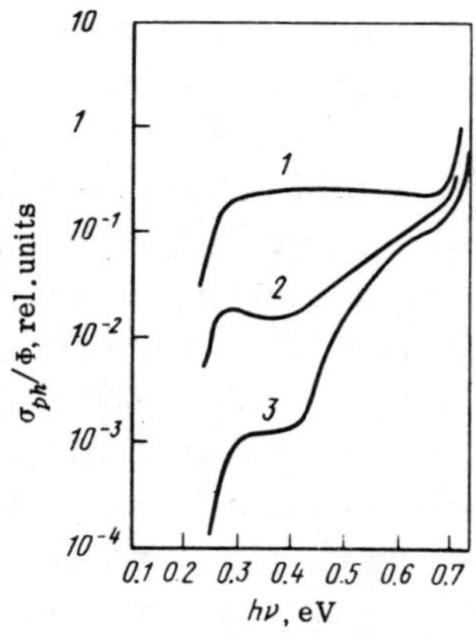

FIG. 65. Photoconductivity spectra recorded at T = 77 K for p-type Ge:Ni samples with the degree of compensation K = 0.06 (1), 0.33 (2), and 0.75 (3).

free electrons because of the $E_t \rightarrow E_c$ transitions. The effective mobility then increased, as indicated by the photo-Hall effect.

The results of a quantitative analysis of the values of $\Delta p/I$ and $\Delta n/I$ were reported in Ref. 175; here, I is the intensity of the impurity-absorbed light; Δp and Δn are the densities of photo-excited holes and electrons, respectively. This analysis showed that the first of these quantities was proportional to $(1 - K)/K$ and the second to the reciprocal $K/(1 - K)$. This provided a reliable confirmation that the two levels were correlated, i.e. that they belonged to the same nickel centre and they differed only in respect of the charge of this centre.

As in the case of Ge:Mn, various heat treatments of Ge:Ni revealed the presence of shallow acceptor levels. These were attributed to the formation of vacancy–donor pairs.

3.2. ENERGY LEVELS OF T-METAL IMPURITIES IN SILICON

A relatively full summary of the ionization energies of T-metal impurities in silicon obtained by various authors, is given in Table XXII. The most reliable values in Tables XXII-XXVIII are identified by asterisks.

Compared with the analogous Table XX for the Ge:T-metal systems, we can see that a number of available values of E_t is large and that they disagree. This is mainly due to the fact that E_t has been rarely found by applying several investigation methods simultaneously. Moreover, because the interactions of point defects creating new associated centres are stronger, silicon is affected by its previous history more than any other semiconductor. These are the effects that influence the values of E_t found for different samples. Finally, we cannot exclude the possibility that some of the T-metal impurities in silicon exhibit a strong Franck–Condon shift because of a strong electron–phonon interaction.

We shall now discuss this point in greater detail. It should be mentioned that even fully reliable determinations of E_t carried out using optical properties do not give values that agree with the thermal ionization energy deduced from the temperature dependences of the resistance R(T) or resistivity $\rho(T)$. In some cases the disagreement is slight and this is probably why it has been ignored for a long time. However, the disagreement between the values of E_t deduced from the photoconductivity and from the Hall effect appears in a regular manner because of the electron–phonon

TABLE XXII. Energy Levels of T-Metal Impurities in Silicon found by Different Authors

Impurity	E_c-E_t, eV	E_v+E_t, eV	Method of determination	Ref.	Impurity	E_c-E_t, eV	E_v+E_t, eV	Method of determination	Ref.
Sc	not investigated				Co	0.39A	0.5A	H	[196]
Ti	0.21	–	MOS	[23]			0.5A	OA	[197]
V	0.41	0.4	MOS	[23]			0.44	H	[198]
Cr	0.22* D	–	H	[13, 186]		0.53 A	0.35A	PH	[199]
	0.4D	–	PS	[190]		0.53*A	0.4D	H	[200]
	0.41D	–	H	[190]		0.52A	0.38D	PH	[203]
	0.23D	0.11D	H	[187]		0.52A	–	OA	[200]
Mn	0.53D	–	H; PS	[188]		0.6A	–	OA	[200]
	0.5*D	–	H; PH; SCA; IQ	[189]		0.73	–	OA; PH	[200]
	0.5	–	PC	[190]		–	0.1A	H	[202]
	0.39	–	SCA	[190]		–	0.37A	H	[202]
	0.21	–	SCA	[190]		0.52A	0.1A	H	[201]
	0.39 D	–	PC; RC	[190]		0.22A	–	H	[201]
	0.53 D	–	PC; RC	[190]		0.57A	0.38D	SCA; RC	[204]
	0.3D	–	H; PH; PHE; IQ	[189]		0.45	0.3	PC	[190]
						0.22	0.36	SCU	[204]
Fe	0.55D	0.4* D	H; PS	[191]	Ni	0.35*A	0.23A	H	[205, 206]
	0.55*A	–	PS	[192]		0.37A	0.24A	H	[207, 208]
	–	0.4D	–	[254]					
	0.29A	–	H	[194]		0.41A	–	OA	[209]
	0.1D	0.4D	C	[195]		0.39A	0.2A	H; C	[208]
	0.14	–	MOS	[23]		0.34A	–	H	[210]

Note. The following notation is used for the methods of determination: C is the temperature dependence of the electrical conductivity; PS is the photoconductivity spectrum; OA is the optical absorption; SCU and SCA are thermally stimulated current and capacitance; RC is the capacitance relaxation; MOS are the capacitance-voltage characteristics of MOS structure; PC is the photocapacitance. The rest of the notation is the same as in Table XX. The asterisk denotes the most reliable value.

interaction[183,184] and can be explained by the scheme in Fig. 66. This figure gives the configurational-coordinate diagram of a centre forming a deep level E_t (found from the thermal data). It is based on the hypothesis of formation of a potential well which may contain an electron bound not only by the impurity atom itself but also by the nearest surrounding atoms. The depth of the potential well and, consequently, the value of E_t depend on the relative position of the centre and of the surrounding atoms. In the first approximation, the position of such a centre can be described by a single coordinate Q, which is known as the configurational coordinate and is plotted along the abscissa in Fig. 66. This coordinate may, for example, be the radial displacement of the surrounding atoms.

The energy of the atoms in the environment is described, for

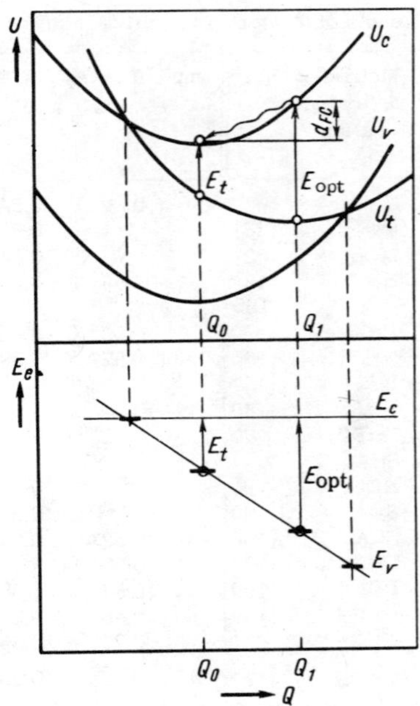

FIG. 66. Configurational coordinate diagram of a centre forming a deep level and characterized by a strong electron–phonon interaction.

the sake of simplicity, by the energy of a harmonic oscillator of mass M vibrating at a frequency ω. Then the energy of the whole system comprising a given impurity centre and its environment can be represented by

$$U = E_e + M\omega^2 Q^2/2. \tag{41}$$

We shall now discuss a donor ionized centre. The vibrations of the atoms do not influence the energy of a free electron in the conduction band and the change in the total energy of the system described by Eq. (41) is entirely due to the second term.

This situation is represented by the curve U_c in Fig. 66.

If an electron undergoes the reverse transition from the conduction band to the level E_t, it follows that – since the depth of the potential well and, consequently, the energy level of a bound electron decrease linearly on increase in Q – Eq. (41) contains both terms which depend in different ways on the coordinate. Hence, it is clear that a system which tends to the minimum energy should occupy a new equilibrium position Q_1 which is different from Q_0.

The value of Q_1 can be found from Eq. (41) subject to the condition $V_t = U_{t\,min}$. The dependence of the total energy of the system in the case of a nonionized donor centre is represented by the curve U_t in Fig. 66. We can now see quite clearly that the Franck–Condon principle makes the thermal ionization energy of the centre E_t different from the optical value E_{opt}.

It follows from the Franck–Condon principle that the optical transition of an electron within a system (molecule) occurs without a change in its configurational coordinate.

The difference between E_t and E_{opt} (Franck–Condon shift) is denoted by d_{FC} in Fig. 66. A similar situation occurs also in the case of a deep acceptor centre. This possibly accounts for the systematic increase in E_{opt} found from optical characteristics, compared with the thermal ionization energies (Table XXII). This effect undoubtedly complicates the problem of investigating the spectrum of deep levels of impurity centres, because a given impurity centre may have two different ionization energies found by thermal and optical methods and this requires an additional analysis.

First of all, we can see from Fig. 66, that an electron returning from the conduction band to a deep centre after the end of photoexcitation has to overcome an activation barrier governed by a relaxation of the electron–phonon system and numerically equal to d_{FC}. This increases the lifetime of a photoexcited carrier. The result is slow relaxation of the photocurrent.

Moreover, participations of the vibrations of the atoms surrounding a given impurity centre is manifested by a change in the law which governs the fall of the photoconductivity (photocurrent) in the long-wavelength part of the spectrum.

In the case of a strong electron–phonon interaction the $j_{ph}(h\nu)$ curve in the region of the red edge is no longer described by Lucovsky law,[159] i.e. not by Eq. (37), but by a different expression[185]:

$$j_{ph}(h\nu) \propto (1/y)[1/(\pi\theta)^{\frac{1}{2}}] \int_1^{\infty} \exp[-(x - y)^2/\theta][(x - 1)^{\frac{1}{2}}/x]dx, \quad (42)$$

where

$$x = E_t/E_{opt};\ y = \hbar\nu/E_{opt};\ \theta = (a\hbar\Omega/E_{opt})^2 \coth(\hbar\Omega/2kT). \quad (43)$$

The quantities in Eq. (43) have the following meaning: a is the dimensionless electron–phonon interaction constant; Ω is the frequency of a phonon interacting with an electron; E_{opt} is the "optical" ionization energy of a T-metal centre.

The quantities a and Ω determine the Franck–Condon shift d_{FC} and the linear shift q of an impurity atom in a crystal lattice as a result of its ionization:

$$d_{FC} = a^2\hbar\Omega/2; \quad q = (a/2)\sqrt{\hbar/m\Omega}, \tag{44}$$

where m is the mass of atoms in the immediate environment.

For convenience in calculations, the dependence (42) is represented in Ref. 185 by a series of normalized curves for different values of θ.

Finally, the Jahn–Teller effect may appear in the course of ionization (or deionization) of T-metal impurities if a T-metal atom acquires an electron configuration with $L \neq 0$. Clearly, the same T-metal impurity can have different energy levels for different electron configurations. This aspect of the energy spectrum has not yet been analyzed for Si:T-metal systems.

We shall now describe briefly the experimental values of E_t obtained for specific impurities.

Scandium

No information is available on the doping of silicon with scandium.

Titanium

There is one report[23] in which the energy level of titanium is given for samples doped by ion implantation. It should be pointed out that ion implantation creates various radiation defects which can interact with the dopant. In spite of the fact that samples were annealed before the measurements,[23] it is not at all clear whether the level in question belongs to a titanium atom in one of the crystallographic positions or whether it represents a titanium atom bound to one of the unannealed radiation defects. Unfortunately, the reported investigation also fails to settle the question of whether the titanium centre is of donor or acceptor nature. Therefore, it is not possible to say anything definite about the electron structure and crystallographic position of titanium in silicon.

Vanadium

The energy level of vanadium in ion-implantation-doped silicon is also reported in Ref. 23. All the doubts relating to Si:Ti apply also to Si:V.

Moreover, an ESR study established reliably[13] that vanadium may exist in silicon in the V^{2+} state (see Sec. 1.3) and have the electron configuration $3d^3$, which corresponds to a vanadium ion in an interstitial position in silicon (Fig. 25).

It was reported in Ref. 13 that these samples had p-type conduction so that the V^{2+} level should be located above the Fermi level. The value E_t given in Table XXII for vanadium in silicon does not conflict with this conclusion.

Unfortunately, there have been no simultaneous investigations of the ESR spectrum and of the temperature dependence of the Hall effect in the same samples of Si:V.

Chromium

A report of a chromium level at $E_c - 0.22$ eV was made in Ref. 13 and was subsequently confirmed by a simultaneous investigation of the Hall effect and the ESR.[186] This level is labelled as the most reliable in Table XXII. It is interesting that all the investigations indicate that the level is a donor. Consequently, we can attribute the level to an interstitial chromium atom with the $3d^5$ electron configuration as found by the ESR method (Table VII).

Table XXII gives also two other energies of a donor level in the region of $E_c - 0.4$ eV. Since there are no reasons to doubt the correctness of this report, the donor level may be attributed to an interstitial but doubly ionized Cr^{2+} (since its configuration is $3d^4$, it cannot be observed by the ESR method) or to Cr^+ at a regular site. The latter identification is less likely because observation of the ESR signal of the site Cr^+ requires a deliberate situation of Si:Cr crystals with vacancies (see Sec.1.3). Another donor level $E_v + 0.11$ eV reported in Ref. 187 is far too deep to be attributed to Cr^+. In our opinion it is more likely to belong to an associated Cr^{2+}–shallow acceptor pair in which Cr^{2+} is a doubly ionized interstitial chromium ion.

Manganese

The energy level of manganese in silicon was first determined to be $E_c - 0.53$ eV from the results of electrical measurements.[188] It was found that the level was a donor.

In addition, there is also a published value of $E_c - 0.5$ eV (Ref. 189). This value was deduced from the results of an investigation of thermal and infrared quenching of the photoconductivity. The same value was obtained in Ref. 189 also from the photostimulated and thermally stimulated conductivity and in Ref. 190 from measurements of the photocapacitance of a Schottky barrier.

Since the position of the donor level of manganese at $E_c - 0.5$ eV was reported repeatedly by many authors using a great variety of methods, it is identified in Table XXII as the most reliable, although in Ref. 177 preference is given to $E_c - 0.53$ eV. Measurements of the photocapacitance of a Schottky barrier[190] and also investigations of capacitance-voltage characteristics of metal–oxide–semiconductor (MOS) structures[23] revealed also other levels (Table XXII). This is possibly due to the fact that manganese (according to ESR data) can have a large number of various states in silicon, including complexes of the Mn_4 type.

Detailed investigations of Si:Mn were carried out also later by Akhmedova and Kamilov.* Akhmedova investigated the photocapacitance and isothermal relaxation of the capacitance (capacitance spectroscopy), whereas Kamilov studied induced photoconductivity, photo-Hall effect, and infrared quenching of the photoconductivity.

The work of Akhmedova is important because it is the only occasion in which the energy spectrum is considered in connection with the cooling regime of a sample after the diffusion doping. This is important because it is the rate of thermal quenching of a sample that determines the state in which manganese atoms are found in silicon. This was demonstrated clearly by the ESR method (see Sec. 1.3).

Rapid quenching prevents the interaction between manganese atoms, whereas slow quenching creates defects of the Mn_4 type. When samples are cooled rapidly, it is found that donor levels appear at $E_c - 0.39$ eV and $E_c - 0.53$ eV, irrespective of the type of conduction before the diffusion of manganese.

Akhmedova attributed the state $Mn^-(3d^8)$ tentatively to the second level, whereas she ascribed the first level to some unidentified complexes containing manganese atoms.

Following this hypothesis, one would expect donor complexes to be most probably Mn^{2+}–shallow acceptor pairs which, as shown in Sec. 1.3, may be manifested in the ESR spectra. However, then the $E_c - 0.53$ eV level should be attributed the Mn^{2+} state rather than to Mn^-. Slow cooling of n-type Si:Mn crystals gives rise to an additional level at $E_c - 0.21$ eV.

The work of Kamilov was equally important. He studied the

---

*M. M. Akhmedova, "Investigation of deep levels in silicon doped with cobalt, platinum, manganese, and copper," Author's abstract of thesis for Candidate's Degree [in Russian], Leningrad Polytechnic Institute (1976).

T. Kamilov, "Investigation of photoelectric properties of manganese-doped silicon," Author's abstract of thesis for Candidate's Degree [in Russian], Leningrad Polytechnic Institute (1977).

evolution of photoelectric properties when the Fermi level shifted from E_v to E_c, which made it possible to demonstrate the existence of an additional donor level at $E_c - 0.3$ eV associated with the presence of manganese in silicon. These results and the data obtained earlier in other but still similar samples by the ESR method make it possible to suggest the following model of the energy spectrum of Si:Mn. The manganese atoms occupy interstitial positions in silicon and are in the neutral state $3d^7$. They can be singly and doubly charged donors (Fig. 25) with the configurations $3d^6$ and $3d^5$. The former corresponds to the energy level $E_c -$ 0.3 eV and the latter to $E_c - 0.5$ eV.

It is difficult to draw any definite conclusions on the nature of the $E_c - 0.39$ eV level. This may be due to manganese at regular sites after a transition from $Mn^0(3d^3)$ to $Mn^+(3d^2)$ or manganese may be in the interstitial position but after accepting an electron, i.e. $Mn^-(3d^8)$. In this case the interstitial manganese manifests amphoteric properties. The level may also be due to Mn–shallow acceptor pairs. This is supported by the observation of the $E_c - 0.39$ eV level mainly in silicon samples exhibiting initially p-type conduction.

The weakness of any of these suggestions is that the value 0.39 eV is also the half-sum of two other values of E_t for manganese, which naturally leads to doubts of whether the 0.39 eV level truly exists.

Finally, the $E_c - 0.21$ eV level observed solely after a slow cooling of n-Si:Mn can clearly be identified with Mn clusters. We cannot exclude the possibility that at certain quenching rates the process of complex formation does not reach the Mn_4 stage, but stops at Mn_3 or Mn_2, and various levels correspond specifically to these stages.

The proposed model does agree with all the experimental data but it obviously must be carefully checked experimentally by simultaneous investigation of electrical and photoelectric properties and of ESR, and better still by a study of photostimulated ESR in the same samples.

Iron

The electrical properties of Si:Fe doped with iron during growth of crystals or by diffusion into n-type and p-type samples were considered in Ref. 191. After diffusion annealing in sealed ampoules at 1200° C followed by quenching in water it was found that p-type samples exhibited donors in a concentration of $1.5 \cdot 10^{16}$ cm^{-3}

and with the energy level at $E_v + 0.4$ eV.

Introduction of iron into n-type Si did not alter the resistivity o the original sample. The acceptor properties of iron were not detected although the samples had a fairly low donor background ($\leqslant 10^{14}$ cm^{-3}).

These two observations confirm the donor nature of the iron impurity in silicon. Naturally, this level should be attributed to the $Fe^{+}(3d^7)$ interstitial state for which the ESR spectrum was observed by other investigators (see Table VII) who used a similar method of diffusion annealing.

An important manifestation of the conversion of the $E_v + 0.4$ eV level to a new level at $E_c - 0.55$ eV was pointed out in Ref. 191. This conversion was fairly rapid (it took several days) even at room temperature.

It was interesting to note that heating to 100–200° C resulted in the opposite conversion of the 0.4 eV level to 0.55 eV. At even higher temperatures it was found that irreversible changes in the properties of Si:Fe took place.

It was concluded in Ref. 191 that the second level at $E_c - 0.55$ eV was also a donor. In Ref. 193 this level was attributed to Fe–shallow acceptor pairs. However, it was shown in Ref. 28 that conversion of the levels left the ESR spectrum of Fe^0 unaffected, whereas pairs exhibited a very different ESR spectrum, demonstrated clearly by a comparison of the data in Tables VII and IX.

It is worth noting a comprehensive investigation of electrical properties and of ESR in the same samples subjected to different heat treatments.[193] The most important result of this investigation was the proof of traces of chromium in silicon samples doped with iron and consequent compensation of both shallow acceptors and shallow donors.

Since the iron level at $E_v + 0.4$ eV is a donor, the compensation of shallow acceptors is self-evident. However, the compensation of shallow donors is possible if iron introduces also an acceptor level in the upper half of the band gap. Therefore, the $E_c - 0.55$ eV level is regarded as an acceptor in Ref. 193.

In view of this information the energy spectrum of levels in Si:Fe can be represented as shown in Fig. 67.

After diffusion annealing and quenching of p-type Si crystals, a shallow acceptor level of boron at $E_v + 0.045$ eV is initially (Fig. 67a) compensated by electrons from the donor level of iron

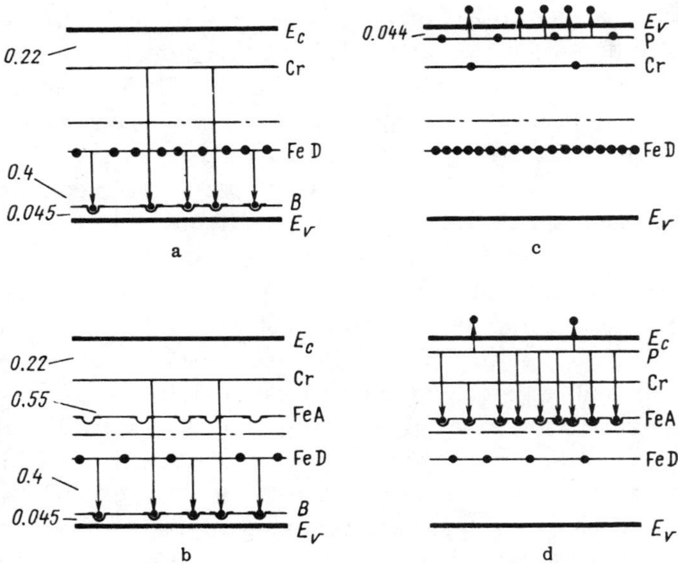

FIG. 67. Model of energy levels in Si:Fe: a), b) p-type silicon; c), d) n-type silicon.

and partly by the donor level of chromium. Therefore, such samples may exhibit ESR due to the remaining neutral Fe^0 atoms. In addition, the ESR spectrum of Cr^+ ions is observed.[193] Storage at room temperature converts the donor level of iron to an acceptor at $E_c - 0.55$ eV (Fig. 67b), which reduces the concentration of Fe^0 and consequently lowers the intensity of the ESR spectrum.

Figure 67c shows the distribution of electrons in n-type Si:Fe after quenching. The concentration of Fe^0 in such a sample is high, but the large number of free electrons from ionized phosphorus levels at temperatures 100-300 K makes the samples so conducting that it is not possible to observe the ESR spectrum of Fe^0 because of a strong fall in the Q-factor of the ESR spectrum at the resonator.

Cooling to 20 K makes it possible to record the ESR spectrum of Fe^0 even in such low-resistivity samples.[13,28]

When samples are stored for a time sufficient to ensure the conversion of the iron levels, the new acceptor at $E - 0.55$ eV collects electrons from the phosphorus and chromium atoms (Fig. 67d). The ESR signal from the remaining interstitial Fe^0 atoms can again be observed. At the same time the ESR lines of the Cr^+ ions appear in the spectrum.[193]

The occurrence of the conversion of the iron levels at room

temperature suggests that the process involves participation of a very mobile defect, which can only be a vacancy.

Therefore, the acceptor level at $E_c - 0.55$ eV may be either due to a site state of iron formed as a result of recombination of an iron atom with a vacancy or due to a pair of the interstitial iron–vacancy type, i.e. due to an $(Fe^{+}–V^{2-})^{-}$ pair. Therefore, doubly charged vacancies must be available in silicon.

It is not yet possible to say anything more definite about the acceptor state of iron in silicon. The model discussed above does not deal with the donor level of iron at $E_c - 0.1$ eV reported in Ref. 191 or the acceptor level at $E_c - 0.29$ eV reported in Ref. 195. It is possible that these energies are simply the half-slopes, which are easy to mistake for a true level. The ability of iron atoms to form regularly associated defects suggests that these levels may be due to some metastable forms of such defects.

These comments apply even more strongly to the level at $E_c - 0.14$ eV deduced from the capacitance-voltage characteristics of an MOS structure,[23] because an iron-doped silicon crystal is subjected to high-temperature treatment during the fabrication of such a structure.

Cobalt

Table XXII makes clear the large discrepancies between the values of E_t attributed to cobalt by different authors.

In one of the first investigations of Si:Co (Ref. 196) the results of electrical measurements were used to find the two levels identified as acceptors with values of E_t lying 0.39 and 0.5 eV above the valence band. The similarity of the behaviour of cobalt and zinc in silicon was noted in Ref. 197 and it was suggested that cobalt could form multiply charged impurity centres. In fact, a study of the optical absorption[198] revealed two peaks of the absorption curve with maxima at 0.495 and 0.354 eV. The same investigation included a determination of the temperature dependence of the density of free carriers which revealed a level at $E_v + 0.44$ eV. It was this level that was attributed in Ref. 198 to cobalt. The absence of investigations of the photoconductivity and of the photo-Hall effect was clearly the reason why the nature of the absorption peaks could not be explained in Ref. 198 and even the nature (donor or acceptor) of the $E_v + 0.44$ eV level could not be identified.

The results of later investigations of the energy spectrum of Si:Co can be divided into two groups. The first consists of Refs.

199 and 200, and the other of Refs. 201 and 202.

In the first group of papers it is concluded that cobalt compensates both n- and p-type Si, i.e. cobalt is an amphoteric impurity. In an earlier report[199] the same authors were of the opinion that cobalt compensated only n-type Si.

Quantitative investigations of $\rho(T)$ and $R(T)$ reported in Ref. 200 gave a donor level in the lower half of the band gap at $E_v + 0.4$ eV and an acceptor level in the upper half at $E_c - 0.53$ eV. The second of these levels was not determined very accurately. The same authors reported also higher values right up to 0.73 eV (Table XXII).

The treatment of cobalt as an amphoteric impurity was based on simultaneous investigations of the photoconductivity and photocapacitance. However, it was concluded in Ref. 200 that cobalt not only forms discrete levels, but also a band of levels near the middle of the band gap of silicon of either type of conduction.

Some of the discrete levels coincide with those reported in the second group of papers[201, 202] where no compensation of p-type Si was reported. Conversely, introduction of cobalt into p-type silicon was found to increase slightly the density of holes. Cobalt in n-type Si reduced strongly the density of electrons and under suitable conditions could result in inversion of the type of conduction.

These observations led to the conclusion[201,202] that all the levels (a total of five) of cobalt in silicon were of the acceptor nature. It seems to us that an acceptor charge in excess of two is unlikely in the case of cobalt ions irrespective of whether these ions are located at substitutional or interstitial positions in silicon. This doubt follows from the possible electron configurations of cobalt shown in Fig. 25.

It also follows from general considerations that the behaviour of cobalt could hardly differ significantly from the behaviour of iron belonging to the same group in the periodic system. If we bear in mind this analogy, we find that the most probable cobalt levels involve a donor at $E_v + 0.4$ eV (or at $E_v + 0.38$ eV) and an acceptor level at $E_c - 0.53$ eV (or at $E_c - 0.52$ eV).

The latter level may, by analogy with iron, belong to the site state Co^-. Since in this state (Fig. 25) a cobalt ion has the $3d^6$ configuration (and not $3d^5$, like Fe^-), it is subject to a strong electron–phonon interaction and consequently the optical and thermal ionization energies of the $Co^-(3d^6)$ may differ considerably. Hence, the values of E_t of this level deduced from the optical data (Table XXII) are high.

As in the case of Si:Fe, the E_c – 0.22 eV level may be attributed to the concomitant chromium impurity captured by samples together with the main (cobalt) dopant.

This model of the behaviour of cobalt must of course be tested by careful investigations. At this stage it provides a consistent explanation of practically all the various values of E_t listed in Table XXII and it agrees with the model of the behaviour of iron in silicon, which is a similar impurity.

Nickel

By analogy with iron, once again we can assume that nickel (at least during the first moments after strong quenching) occupies the Si lattice interstices. It should behave as a donor in this state (Fig. 25). However, none of the investigations have revealed the presence of donor levels of nickel. Conversely, the temperature dependences of the carrier density[205,206] revealed clearly two acceptor levels at E_v + 0.23 eV and E_c – 0.35 eV. Almost all these values were confirmed by later investigations.[207,208]

Careful measurements of the Hall coefficient and knowledge of the total concentration of labelled nickel in silicon[208] made it possible to identify reliably the observed levels with singly and doubly charged nickel ions, respectively. Although the experiments reported in Ref. 209 revealed the presence of just one level in the upper half of the band gap, it was quite clear that the lower half should also have a second acceptor level, because otherwise there would be no change in the type of conduction of the original n-type Si which was reported when nickel was doped by diffusion.[209]

A comparison of the acceptor nature of nickel levels and their double charge with possible electron configurations (Fig. 25) resulted in the only possible conclusion that nickel atoms occupy sites in the crystal lattice of silicon, which was also indicated by the ESR spectra (Sec. 1.3).

We shall now try to give a systematic account of energy levels of T-metal impurities in silicon and germanium. We shall do this by reducing the most reliable values of E_t in Table XXII to the same reference system allowing for their donor or acceptor nature.

We shall also see how the ionization energy of a level varies with the configuration. This is shown in Fig. 68.

We shall first consider the relationship obtained for Si:T-metal. Almost all the T-metal impurities, with the exception of nickel in the site state Ni_s^-, fit the curve drawn through the experimental points. The result for nickel is off the curve because in the

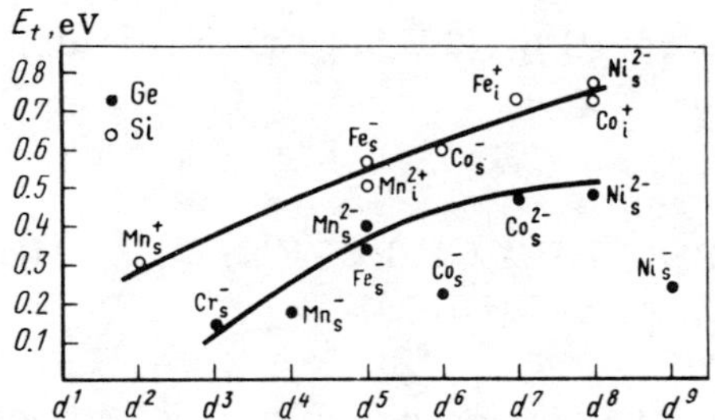

FIG. 68. Dependence of the ionization energy on the occupancy of the d-shell of T-metal impurities in germanium and silicon.

substitutional (site) state it is subject to the Jahn–Teller effect (Sec. 1.3). In the absence of this effect the singly charged Ni_s^- ion would have had the $3d^7$ configuration and the ionization energy $E_t \approx 0.7$ eV (as deduced from the curve in Fig. 68). In fact, it is in the $3d^9$ state and its ionization energy is 0.23 eV. The difference between the energies $\Delta E = 0.47$ eV should be the so-called Jahn–Teller energy which is the energy of deformation of the crystal lattice because of its distortion by the displacement of a nickel ion from a site.

The attachment of a second electron induces the nickel state Ni_s^{2-} and the Jahn–Teller effect is no longer active, so that the ion acquires the electron configuration $3d^8$. Therefore, the experimental value of E_t for Ni^{2-} fits the curve in Fig. 68.

This curve makes it possible to identify the value $E_t = 0.3$ eV found for the manganese donors (Table XXII) with the only possible configuration $3d^2$, i.e. with the state Mn_s^+, which had not been known before.

The data for Ge:T-metal are somewhat poorer but nevertheless they also fit a similar dependence. The above analysis applies also to both Ni_s^- and Ni_s^{2-} ions in germanium. The point representing the donor state of cobalt, identified by Co_i^+ in Fig. 68, should be regarded as an associated defect which contains cobalt. This was the interpretation adopted in Ref. 179.

The situation is different in the case of Co_s^- in the $3d^6$ state. The fact that it is off the curve for Ge:T-metal may be attributed to the Jahn–Teller effect, but then we cannot see why the same does not apply to Si:Co. Therefore, it is not possible to account for the position of this point off the curve.

Plotting of the values of E_t for Ge:Cr made it possible to identify the chromium level with the $3d^3$ electron configuration. It therefore follows from Fig. 25 that the chromium level represents the Cr_s^- state.

3.3. ENERGY SPECTRUM OF LEVELS OF T-METAL IMPURITIES IN GALLIUM ARSENIDE

Many investigations have been made in order to determine experimentally the positions of energy levels of transition metal impurities in gallium arsenide. Reviews of the published data can be found in Refs. 177 and 211. The results of all investigations (without exception) demonstrate that doping of gallium arsenide with T-metal impurities generates deep acceptor levels. Their presence is revealed by studies of the temperature dependences of equilibrium electrophysical parameters and also as a result of investigations of optical and photoelectric properties of GaAs:T-metal crystals. In contrast to germanium and silicon, much information on GaAs:T-metal samples can be obtained by studies of luminescence.

One of the first determinations of the energy spectrum of gallium arsenide doped with transition metals was reported by Haisty and Cronin[76] who determined the temperature dependences of the electrical resistivity and of the Hall coefficient of gallium arsenide single crystals doped with chromium, iron, nickel, cobalt, and manganese. The dependences obtained by them are plotted in Fig. 69.

Other authors subsequently obtained similar results. We shall now consider separately each of the T-metal impurities in gallium arsenide in the order in which they were investigated.

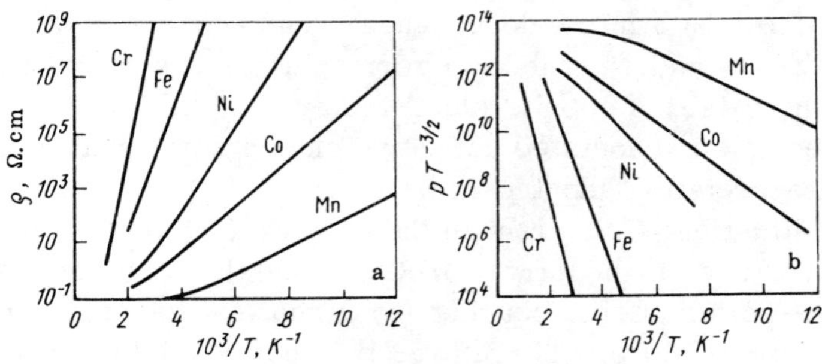

FIG. 69. Temperature dependences of the electrical resistivity (a) and of the hole density (b) in gallium arsenide doped with transition metals.[76]

TABLE XXIII. Energy Levels of Iron in Gallium Arsenide

$E_v + E_t$, eV	Method	Ref.	$E_v + E_t$, eV	Method	Ref.
0.37	C	[212]	0.59; 0.36	TS	[220]
0.52	H	[76; 213; 214]	0.22	PL	[221]
0.21	H	[216]	0.21	EL	[216]
0.58; 0.38	H	[217]	0.36; 0.43	EL	[222]
0.43–0.47	C; H	[76]	0.5	CL	[223]
0.55	OA	[218]	0.5; 0.8	H; OA; ESR	[224, 225]
0.5; 0.37; 0.8	PH	[219]	0.49*; 0.64	H; OA; PH; ESR; PAR; PESR	[47]
0.41	PH	[214]			

Note. The notation used for the method of determination of the energy levels is as follows: C denotes the temperature dependence of the electrical conductivity; H denotes the temperature dependence of the Hall effect; OA denotes the optical absorption method; PH is the impurity photoconductivity; TS is the tunnel spectroscopy; PL, EL, and CL are the photoluminescence, electroluminescence, and cathodoluminescence, respectively; ESR denotes the electron spin resonance method; PAR is the paramagnetic acoustic resonance; PESR is the photoinduced electron spin resonance.

Iron

Table XXIII gives a complete summary of the values of E_t deduced by various authors. The iron impurity was first identified with a level at $E_t + 0.37$ eV deduced from the temperature dependence of the electrical resistivity.[212] The reliability of this result can be judged on the basis of an equation valid in the case of one deep acceptor level: $p = [(N_1/N_2) - 1]p_1$. The quantities N_1 and N_d represent the concentrations of the acceptors and of the donor background, respectively,

$$p_1 = (N_v/g_1)\exp(-E_1/kT). \tag{45}$$

An elementary calculation and substitution of $E_t = 0.37$ eV together with the most probable values $g = 4$ and $N_d \sim 10^{16}$ cm^{-3} gives a concentration of deep centres $2 \cdot 10^{19}$ cm^{-3}, which is more than an order of magnitude higher than the solubility limit of iron in gallium arsenide. Clearly, the 0.37 eV level cannot be due to iron in the state of substitution for gallium at the regular sites, such substitution having been established reliably from the ESR spectra (Sec. 1.4). We can only assume that at the time when the work reported in Ref. 212 was carried out the level of the gallium arsenide technology was not sufficiently high and the doping with iron resulted in generation of defects with a level at 0.37 eV either due to structural imperfections or due to the interaction of the dopant with other point defects. The latter assumption is in agreement with the results of other investigations where the same or similar level was reported.[217,219,220] For example, according to Refs. 219 and 220, iron was captured during the process of formation of a p-n junction when impurity interactions are more probable.

The detailed data (Fig. 69) were attributed by the authors to a deep level at $E_v + 0.52$ eV. This value was obtained by many authors irrespective of the method used to introduce iron into a crystal (growth from the melt, diffusion, epitaxy).[214,215,226]

The temperature dependence of the ionization energy of a deep iron level was reported in Ref. 213: $E_t = E_t(0) + \alpha T$, where α is a constant.

The obvious problem is related to the temperature dependence of the band gap E_G of the host matrix. In contrast to classical hydrogen-like levels linked quite firmly to the absolute extremum of the conduction band or to the top of the valence band, deep levels cannot strictly speaking be regarded as associated with any particular extremum or even with one energy band (Chap. 2). Since the energy separation between a deep level and many extrema $E(\vec{k})$ are quantities of the same order of magnitude, the wave function of a deep centre is a superposition of all the states in both allowed energy bands. Therefore, when $E_G(T)$ varies, a deep level cannot be regarded as "linked" to one energy band. A theoretical calculation of the dependence $E_t(T)$ for deep centres is a problem still remaining to be solved. Therefore, the required dependence was found in Ref. 213 from the experimental results.

The value of α was found in Ref. 213 from the measurements of $\rho(T)$ right up to the impurity exhaustion region, but the conversion of the results to the temperature dependence of the hole density $p(T)$ was not quite correct because the ambipolar nature of the equilibrium conduction process begins to manifest itself at $T > 500$ K. Extrapolation of the temperature dependence of the hole mobility $u_p(T)$ to the high temperature range can also give rise to significant errors. On the other hand, the influence of the dependence $E_t(T)$ on $p(T)$ is weak and it manifests itself only at $T > 500$ K. Therefore, the error associated with these approximations and the unreliability of the selected values of the g-factor and of the compensating donor background N_d lead to doubts about the correctness of the conclusion that the temperature dependence $E_t(T)$ is strong (and that the constant is $\alpha = -4.5 \cdot 10^{-4}$ eV/K), as a result of which a deep acceptor level of iron is strongly linked to the bottom of the conduction band.[213]

Measurements of the equilibrium effect cannot be used to determine accurately the value of α even if the concentrations N_a and N_d are known accurately, because of the indeterminacy of the g-factor. The dependences $p(T)$ are more likely to give the effective degeneracy factor $g^* = g\exp(-\alpha/k)$, which together with the degree of compensation N_d/N_a governs the ordinate of the dependence $\log pT^{-3/2} = f(10^3/T)$. We can then expect α to represent only a fraction of the change in the band gap $(d/dT)E_G$.

The method of the temperature dependence of the electrical resistivity and of the Hall coefficient had been used to determine the ionization energy of the iron level also in other investigations.[215–217] In Ref. 217 the slope of the curve representing the Hall effect was used to find the value $E_t = 0.58$ eV. In Ref. 227 an analysis was made of crystals containing iron in concentrations in excess of 10^{18} cm^{-3} for which the values of E_t deduced from the Hall effect varied within the range 0.43-0.47 eV. It was not clear whether these values were due to some variations in the properties of the GaAs:Fe system at impurity concentrations in excess of the maximum solubility or whether they were the result of generation of defects with the ionization energy close to that of iron. The value of $E_t = 0.21$ eV attributed in Ref. 216 to the iron impurity was even more difficult to account for.

Most likely this last value was not representative of iron but of a defect introduced during the diffusion process employed in the preparation of the samples. It should be pointed out that in all investigations with the exception of those reported in Refs. 47 and 213, the equilibrium values of the electrical resistivity ρ and of the hole density p were determined within a narrow range of temperatures because of methodological difficulties and this could have given rise to considerable errors in the values of E_t.

In fact, the results reported in Ref. 47 indicated that in a fairly narrow temperature interval the slope of the dependences $\ln pT^{-3/2} = f(10^3/T)$ obtained for samples with different iron concentrations was the same and amounted to 0.49 ± 0.01 eV.

However, the slope of the Hall effect curves changed somewhat at higher temperatures. Beginning from ~250 K, the slope rose slowly to 0.58 eV and the range of such gradual increase in the slope extended over four orders of magnitude in respect of the hole density.

An analysis of these results showed that they represented a semiconductor with energy levels split by the crystal field.

The lowest of the values of E_t amounting to $E_v + 0.49$ eV can be identified with the ground state of an ionized iron impurity. Moreover, investigations of ESR and of photo-induced ESR of the same samples[47,225] made it possible to link unambiguously this level to a transition from the $Fe^{3+}(3d^5)$ state to the state Fe^{2+} with the electron configuration $3d^6$, i.e. to an iron impurity atom located at a Ga site in the crystal lattice of gallium arsenide after capture of an electron from the valence band. The second deeper level at $E_v + 0.64$ eV, found for the same samples in Ref. 47, could be identified in an obvious manner with the split-off level of the same Fe^{2+} centre. However, the energy gap between them amounting to 0.16 eV is approximately half the diameter representing splitting by the crystal field $\Delta = 0.37$ eV and is much greater than splitting λ by the spin–orbit interaction (see Sec. 1.4).

It is possible that the value $E_v + 0.64$ eV was determined incorrectly because of the difficulty in estimating the g-factor of this level. For these reasons it is not possible to identify the most reliable energy of an excited (split-off) level of iron in gallium arsenide.

The $^5T_2$ excited state of the Fe^{2+} ion observed in the absorption spectrum cannot be manifested in the dependences p(T) or ρ(T). This is due to the fact that the state in question is located above the mid-point of the band gap and that release of intrinsic equilibrium carriers (holes and electrons) occurs before filling of the $^5T_2$ level by electrons from the valence band.

Figure 70 shows the absorption spectra recorded at 300 K for GaAs:Fe samples with different concentrations of the dopant.[224] It was found that the absorption coefficient measured at a fixed photon energy was directly proportional to the concentration of the deep impurity in the Fe^{3+} configuration, found from the ESR data,

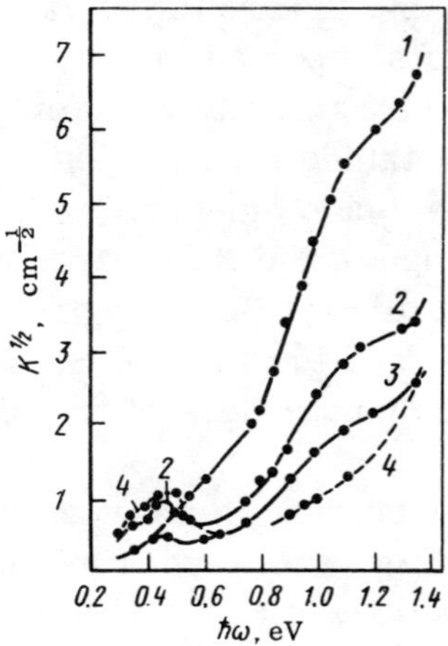

FIG. 70. Absorption spectra of GaAs:Fe at 300 K: 1), 3) samples with the accidental background of compensating donors; 2), 4) samples doped simultaneously with iron and tellurium (curve 2 represents a p-type sample and curve 4 represents a low-resistivity n-type sample).

with the exception of the range of concentrations $N(Fe^{3+}) \leqslant 10^{16}$ cm^{-3}. It follows that the absorption in the impurity part of the spectrum is associated with the photoionization of neutral iron Fe^0, i.e. with optical release of electrons from the valence band to the iron level.

These spectra exhibited a feature manifested by a kink of the curves in the region of 0.8 eV. It can be explained by assuming that, in addition to photoionization of deep iron levels, transitions from the valence band to states located near the mid-point of the band gap begin to manifest themselves at this energy. If the concentration of the deep-level centres is in no way related to the concentration of iron in the investigated material, the contribution of these centres to the absorption will be strongest in samples with low Fe^0 concentrations. This hypothesis is well supported by the experimental results.[228]

An analysis of the dependences $\alpha(h\nu)$ for samples with low iron concentrations ($\sim 2 \cdot 10^{15}$ cm^{-3}) carried out using Eq. (37) shows a good agreement with the Lucovsky model and it makes it possible to find the ionization energy $E_t + 0.8$ eV and the concentration $(4-8) \cdot 10^{16}$ cm^{-3} of deep centres responsible for the shorter-wavelength absorption band. Both these parameters agree with those of the oxygen impurity in the investigated samples.[228] Therefore, the 0.8 eV level is in no way related to the presence of iron in gallium arsenide.

The curves in Fig. 70 have another singularity: there is a weak structure in the region 0.3-0.5 eV. It is resolved by cooling to 4.2-20 K (Fig. 33) and makes it possible to determine the full set of split-off terms of the Fe^{2+} ion (Fig. 34), as discussed in Sec. 1.4.

Determination of E_t of the iron impurity from the photoconductivity spectra is even more difficult than from the optical absorption spectra. It was reported in Ref. 220 that for the majority of GaAs:Fe samples the properties were not affected by illumination, i.e. there was no significant photomemory. However, in some cases a higher photosensitivity was observed in the region of 1 eV. After short-wavelength illumination these samples could remain in a nonequilibrium state for a considerable time indicating that fairly deep traps participated in the charge exchange processes. When illumination was switched off, the conductivity of these crystals was much higher than the equilibrium value. All these features were typical of crystals with a low concentration and a strong compensation of deep iron centres.

Figure 71 shows typical photoconductivity spectra obtained at various temperatures for crystals without the photomemory effect. This figure includes also the dependence (dashed curve) calculated using the Lucovsky model which describes quite satisfactorily the long-wavelength edge of the photoconductivity spectrum. The results plotted in Fig. 71 indicate some shift of the photoconductivity edge in the direction of lower energies as temperature is increased.

However, we cannot deduce from these data that the temperature dependences of E_t and E_G are identical because of the thermal smearing of the edge by an amount representing a few kT.

The spectra in Fig. 71 demonstrate an increase in the photosensitivity of GaAs:Fe samples on increase in temperature which can be explained by the temperature dependence of the cross-section $\sigma_p(T)$ for the capture of nonequilibrium holes by negatively charged Fe^- centres. It is therefore necessary to assume that σ_p decreases on increase with temperature as $\exp(0.04/kT)$ or T^{-4}. Such a strong temperature dependence of the capture cross-section is difficult to account for by the existing theories although such dependences $\sigma_p(T)$ have been observed earlier for other semiconductor–impurity systems. For example, an effect of this kind in Si:In is attributed to the trapping of holes by excited indium states.

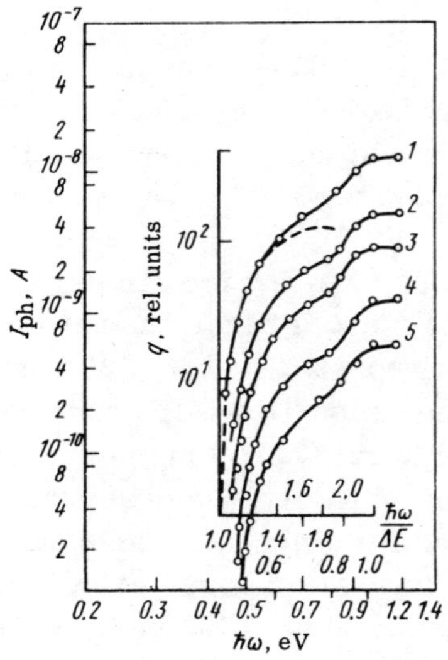

FIG. 71. Photoconductivity spectra of GaAs:Fe samples at the various temperatures (K): 1) 210; 2) 192; 3) 155; 4) 130; 5) 102. The dashed curve is the spectral dependence of the photoionization cross-section.[60]

However, the strong temperature dependence of the photoconductivity of GaAs:Fe samples can also be explained by the existence of shallow hole traps of energy ~0.04 eV. In the case when the hole-capture cross-section is $\sigma_t^p > \sigma_{Fe}^p$ and the trap concentrations obey $N_t \gtrsim N_{Fe}^-$, these levels can affect significantly the low-temperature occupancy of the iron centres which govern the non-equilibrium hole lifetime: $\tau_p = (1/v)\sigma_{Fe}^p N_{Fe}^-$. It should be noted that hole traps with the ionization energy 0.03-0.04 eV were found in studies of the intrinsic photoconductivity of GaAs:Fe in the same temperature range.[215]

Both these explanations are linked in one way or another to the trapping processes. In the former explanation the traps are the excited states of deep centres, the existence of which has not yet been proved.

We can also expect that if the excited states of deep centres do exist, that they should be hydrogen-like and have a low ionization energy. In fact, at large distances from an impurity an electron experiencing the Coulomb attraction by a deep centre may form bound states the energies of which are governed by the effective Rydberg constant $E_n = R^*/n^2$ and differ little from the excited states of shallow impurity centres.

According to the latter explanation the trapping levels are independent centres and may be associated with such impurities as copper (E_v + 0.023 eV) or silicon (E_v + 0.03 eV). In either case the origin of the traps is not clear and further studies are needed to determine it more reliably. Such studies can be justified by the search for excited states of deep centres in GaAs:Fe which are of intrinsic interest.

Table XXIII lists additional values of E_t found by the luminescence methods. Those which were determined by a study of the photoluminescence and electroluminescence can hardly be associated directly with iron because the iron levels are involved mainly in nonradiative recombination.

In studies of cathodoluminescence the excitation rate is usually so high that even a small contribution of radiative recombination becomes fully observable. A cathodoluminescence consisting of two fairly wide bands with photon energies 0.32 ± 0.02 and 0.5 ± 0.02 eV was reported in Ref. 229. The long-wavelength transition at 0.32 eV appeared at a relatively low density of the excitation current and it grew rapidly. At higher currents the short-wavelength peak at 0.5 eV predominated. Within the limits of the experimental error, the energies of both maxima were independent of the excitation rate. When the temperature of a sample was increased to 300 K, the 0.32 eV band disappeared; only the band at 0.5 eV was observed and its energy position remained unaffected.

These two cathodoluminescence bands can be attributed in a natural manner to radiative transitions involving the participation of iron impurity levels. The former with the energy 0.32 ± 0.02 eV is clearly due to intracentre transitions between the $^5T_2$ and ^5E states of Fe^{2+}.

Some decrease in the energy of this band compared with the splitting by the crystal field $\Delta = 0.37$ eV is due to the fact that at 77 K the bulk of radiative transitions are phonon-assisted.

The second cathodoluminescence band at 0.5 eV is probably due to the capture of nonequilibrium holes by a deep iron level, which governs both equilibrium and nonequilibrium properties of semi-insulating GaAs:Fe.

It is important to stress that, according to the results obtained, the energy position of this level is constant to within ±0.02 eV at temperatures in the range 90–300 K. This is in agreement with our conclusion that the results of Ref. 213 on $E_t(T)$, reported above, are incorrect.

In this way the value of E_t amounting to $E_v + 0.49$ eV has been confirmed reliably also by nonequilibrium phenomena. This is why it is identified in Table XXIII as the most reliable (it is denoted by an asterisk) and representing the ground state of the Fe^{2+} acceptor ion at the position of the gallium site in the crystal lattice of GaAs.

Chromium

The ionization energies of the chromium impurity in GaAs obtained by various authors are listed in Table XXIV. The results show that chromium atoms in GaAs form very deep local levels located near the mid-point of the band gap. Consequently, the equilibrium parameters of GaAs:Cr differ little from the corresponding parameters of an intrinsic material and the investigated samples frequently exhibit discrepancies in respect of the type of conduction deduced from the thermoelectric power or from the Hall effect. This discrepancy is evidence of the ambipolar nature of conduction, i.e. of the presence of both electrons and holes in approximately the same densities.

However, many authors have calculated n employing the simple relationship $n = (eR)^{-1}$, which undoubtedly can result not

TABLE XXIV. Energy Levels of Chromium in Gallium Arsenide

$E_c - E_t$, eV	Method of determination	Ref.	$E_c - E_t$, eV	Method of determination	Ref.
0.79	H	[76, 230]	0.82	PH	[237]
0.85	H	[231]	0.86; 0.64	PH	[238]
0.80*	H; PL; CL	[214, 232]	0.87; 0.68	PH; EA	[239]
0.79	OA	[230, 233]	0.80	PL	[240]
0.82	OA	[59]	0.7; 0.8	PL	[234]
0.75	PH	[214, 234, 235]	0.8	CL	[223]
0.78	PH	[233]	0.72; 0.82	OA; SCU; PL	[227]
0.79	PH	[236]			

Note. The notation used for the methods of determination of the energy levels is as follows: H denotes the temperature dependence of the Hall effect; PL, CL are the photoluminescence and cathodoluminescence, respectively; OA is the optical absorption; PH is the impurity photoconductivity; EA is the electroabsorption; SCU is the thermally stimulated current method.

only in significant (up to two orders of magnitude) errors in the determination of the absolute values of the carrier densities, but also in a considerable scatter of the values of E_t deduced from the Hall effect data for various samples.

The ambipolar nature of conduction in GaAs:Cr has been allowed for in some investigations. A method with the smallest number of the fitting parameters was used in the work reported in Ref. 232.

This investigation was concerned with two groups of samples. The first group comprised crystals doped only with chromium. Deep acceptors were compensated by the background of residual donors N_d which governed the properties of the undoped material. Since the value of N_d was less than $1.5 \cdot 10^{16}$ cm^{-3}, the acceptor centres of chromium in samples of this group were mainly in the neutral state.

In contrast to crystals of the first group, those belonging to the second group were doped not only with chromium, but also with tin impurities, which created a definite and controllable concentration of shallow compensating donors. The concentrations of these impurities were determined by the radioactive tracer method because both impurities contained the radioactive isotopes (^{113}Sn and ^{51}Cr).

An analysis of the experimental results showed that the equilibrium parameters of the samples belonging to the first group were governed by the ambipolar nature of conduction. On the other hand, the influence of free holes was weak in samples of the second group and the electron density was calculated quite justifiably from the simple relationship $R = (en)^{-1}$.

The dependences n(T) found for samples of both groups were

represented in the form $\ln nT^{-3/2} = f(10^3/T)$ and they had practically the same slope amounting to 0.81 ± 0.01 eV. This was the energy gap $E_c - E_t$. The position of this level was determined in Ref. 232 also relative to E_v by plotting the dependence $\log pT^{-3/2} = f(10^3/T)$, where $p = n_i^2/n$.

It was found that in the latter case the slope of the straight lines was $E_t - E_v = 0.79 \pm 0.01$ eV. Therefore, $E_t = 0.8$ eV seemed to be the most reliable value of the ionization energy of the chromium impurity in gallium arsenide.

It is clear from Table XXIV that much work has been done on the optical and photoelectric properties of GaAs:Cr. However, the majority of the investigations have been concerned with crystals with an unknown degree of compensation of the chromium levels and this is clearly the main reason for the discrepancies between the reported results. In fact, since E_t of chromium is located in the middle of the band gap, we can expect double optical transitions and such transitions make it difficult to determine the position of the chromium level.

The interpretation of the experimental results is complicated also by the fact that the recombination of carriers in GaAs:Cr may occur not only at the chromium level but also in a large number of additional centres present in concentrations comparable with that of chromium.

Nevertheless, investigations of the photoconductivity, photo-Hall effect, and optical absorption carried out together with measurements of the equilibrium parameters on the same samples with known values of N_{Cr} and N_d are very valuable for a reliable determination of E_t.

Such comprehensive investigations were reported in Refs. 59, 241, and 242.

The optical absorption spectra of GaAs:Cr reported in Ref. 59 were already given by us earlier (Figs. 36 and 37), whereas typical impurity photoconductivity spectra from Ref. 241 are reproduced in Fig. 72. They represent GaAs:Cr samples with different degrees of occupancy of the deep levels. The photoconductivity spectrum of a weakly compensated crystal (curve 1) was similar to the absorption spectrum of the same material (Fig. 36) and was due to optical transitions from the valence band to the local states of chromium. In fact, measurements of the photo-Hall effect reported in Ref. 241 showed that in the impurity part of the spectrum the photocarriers were holes.

However, on increase in the degree of compensation, the

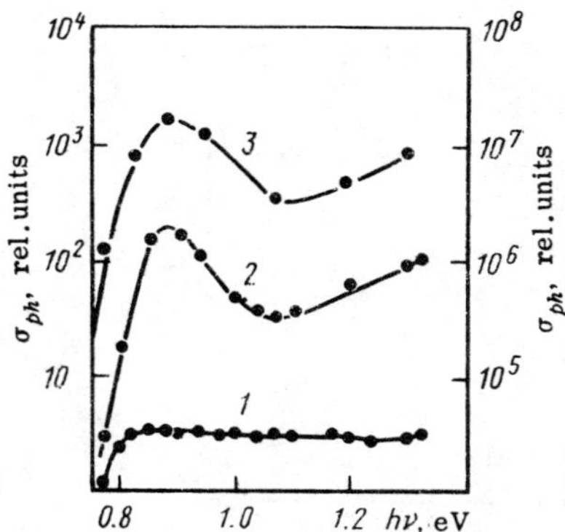

FIG. 72. Impurity photoconductivity spectra of GaAs:Cr at 77 K for different degrees of compensation $k = N_d/N_a$: 1) 0.04; 2) 0.8; 3) 0.98. The scale for curve 3 is given by the right-hand ordinate.

nature of the photoconductivity spectra of GaAs:Cr samples belonging to the second group changed significantly. In the region of $h\nu \sim 0.9$ eV a resonance photoconductivity band was observed. The position of the maximum of this band agreed with the energy of an absorption peak due to intracentre transitions between the ground $^5T_2$ state of Cr^{2+} and the excited ^5E state, which were in resonance with the continuum. The majority carriers deduced from the photo-Hall effect were electrons. The long-wavelength photoconductivity edge, which should be identified with the optical energy of a deep chromium level, was the same for both samples and amounted to ~0.75 eV.

A further increase in the degree of compensation of the chromium levels ($k \to 1$) did not alter the nature of the photoconductivity spectrum, but the photosensitivity increased strongly. This was a manifestation of large-scale fluctuations of the impurity potential in strongly compensated semiconductors and the appearance of a recombination barrier for photocarriers, which increased strongly the effective lifetime, as shown in Ref. 243.

In some cases (Table XXIII) the energy spectrum of GaAs:Cr samples was investigated by the luminescence methods. The photoluminescence spectrum included a characteristic structure-free band with its maximum at 0.8 eV. This band has been attributed to the capture of an electron[214,234,240,244] or of a hole[245] by the chromium level.

In addition to this band, the photoluminescence spectrum had another wide band at 0.6 eV. Figure 73 shows the photoluminescence, optical absorption, and photoconductivity spectra taken from

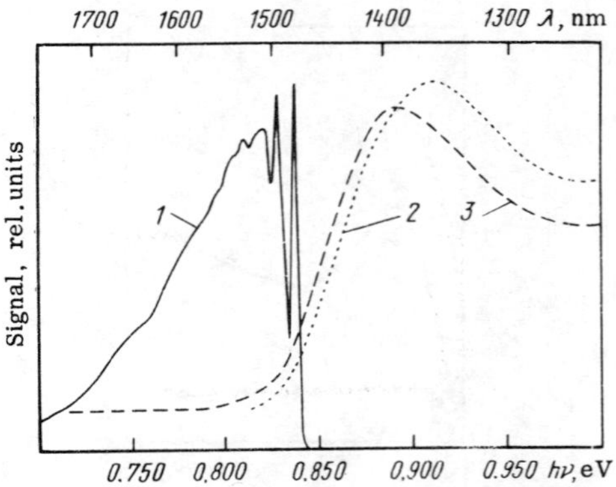

FIG. 73. Photoluminescence (1), optical absorption (2), and photoconductivity (3) spectra of GaAs samples at 4.2 K (Ref. 246).

Ref. 246, where the measurements were carried out at 4.2 K. It is clear from this figure that the zero-phonon line at 0.838 eV and the whole photoluminescence spectrum are shifted strongly relative to the optical absorption and photoconductivity spectra.

Two explanations of these observations have been put forward. The first is related to the position of the excited chromium state ^5E in the conduction band (see Sec. 1.4). In fact, a nonequilibrium electron transferred as a result of optical excitation high into the conduction band ($h\nu > E_G$) tends to pass its excess energy to the lattice by collisions and to occupy a state near the bottom of the conduction band. During such rapid (~10^{-13} sec) thermalization this electron passes through a part of the continuum excited by mixing with the impurity term ^5E. The resonance hardly alters the density of states in the conduction band. We can show that in the region of the resonance the density of states rises by just ~5%. However, the main influence of the hybridization is manifested by the fact that there is a local resonance-like increase in the probability of impurity recombination at $^5T_2$ via the intracentre channel $^5E \rightarrow$ $\rightarrow {}^5T_2$, although the ^5E state is "deformed" by the interaction. The rate of such intracentre transitions is governed by the ratio of the duration τ_c of the capture of an electron by the $^5T_2$ state to the thermalization time τ_E. Obviously, $\tau_E \ll \tau_c$, so that the impurity photoluminescence of GaAs:Cr should be weak. However, radiative transitions from the s-like bottom of the conduction band to the d-states of the chromium impurity are also unlikely. Therefore, mixing of the ^5E state with the continuum is a favourable factor for the manifestation of the impurity photoluminescence of GaAs:Cr. The band singularity responsible for the zero-phonon photoluminescence line at 0.838 eV can be found on the basis of a detailed theory of the resonance interaction with various band states, which is still lacking. However, we can easily see that whereas the photoluminescence should be mainly due to the lower of the hybridized states, all the perturbed part of the conduction band should be manifested in the absorption. The experimental results on the impurity photoluminescence could thus be explained also by intracentre transitions between the ^5E and $^5T_2$ states of the Cr^{2+} ion located in the tetrahedral field of the matrix.

In this explanation the 0.6 eV band in the photoluminescence spectrum is attributed to a recombination centre not associated with chromium but more likely to be due to oxygen. This is in agreement with the observation of such a photoluminescence band in GaAs:O (Ref. 247).

A different explanation of the 0.8 eV band (Fig. 73) is put forward in Ref. 246, where it is assumed that the band in question is due to transitions from the bottom of the conduction band to the $^5T_2$ impurity state and the intracentre $^5E \rightarrow {^5T_2}$ transitions are identified with the 0.6 eV band (in this case the excited ^5E term is assumed to be within the band gap and not in resonance with the continuum). An argument used in support of this explanation is the moderately large splitting of the terms by the crystal field, which would have to be large in the first explanation.

It follows that the position of the split-off Cr^{2+} level cannot yet be regarded as reliably determined, in contrast to the ground state at $E_c - 0.8$ eV.

Nickel

The first report that nickel creates deep traps in GaAs was given in Ref. 248. A large number of papers on nickel impurities has been published subsequently. Some of the results reported in these papers are given in Table XXV. It is clear from this table that the levels at 0.42, 0.35, and 0.2 eV are observed most frequently. The 0.42 eV level has been deduced from the temperature dependences of the electrical conductivity and Hall coefficient of GaAs:Ni samples with a room-temperature conductivity $3 \cdot 10^2\ \Omega.cm$ and a carrier density 10^{12}–10^{14} cm^{-3}. A level with a similar energy of 0.44 eV has been deduced from the cathodoluminescence spectra.[223]

Observation of a deep level at 0.4 eV in GaAs:Ni samples was reported in Ref. 249, where the temperature dependences of the Hall coefficient and optical absorption were studied. The same investigation revealed a level at 0.2 eV and this was attributed to

TABLE XXV. Energy Levels of Nickel in Gallium Arsenide

$E_v + E_t$, eV	Method	Ref.	$E_v + E_t$, eV	Method	Ref.
0.22	H	[81]	0.22	PH	[253]
0.2; 0.44	H	[249]	0.21; 0.33–0.43	PL	[252]
0.2; 0.35	H	[251]	0.35	EL	[222]
0.42	C; H	[250]	0.54	CL	[223]
0.42	OA	[220]	0.53	TS	[254]
0.21	PH	[76]	0.35*	H; PH; OA; ESR	[75, 253]
0.21; 0.38	PH	[214]			

Note. The notation is the same as in Tables XXIII and XXIV.

the formation of structure defects in GaAs:Ni. A level with this activation energy had been observed also in GaAs:Fe and in undoped GaAs.

The temperature dependence of the Hall coefficient was used in Ref. 214 to deduce the presence of levels at 0.38 eV in single crystals grown from the melt and at 0.21 eV in epitaxial films.

An investigation of the temperature dependence of the Hall coefficient of GaAs:Ni was also reported in Ref. 251. Calculations were used to determine the most probable value of the activation energy of a deep impurity centre (0.2 eV), also of the concentrations of the deep-level centres and of the compensating donors, and of the value of the degeneracy factor. A theoretical dependence $p = f(10^3/T)$ given in this paper was in quite satisfactory agreement with the high-temperature experimental data. However, at low temperatures the experimental results deviated from the theoretical curve. This deviation was attributed to the formation of an impurity energy band because of the high impurity concentration. This seemed to be unlikely because the solubilities of transition-metal impurities in GaAs are low[76] and it is difficult to envisage the formation of an impurity energy band. The impurity photoconductivity spectra of GaAs:Ni reported in Ref. 251 revealed two levels with the ionization energies ~0.25 eV, in agreement with the level deduced from the Hall effect measurements, and 0.35 eV, attributed tentatively to nickel.

These values were close to those deduced from the photoluminescence spectra.[252] The room-temperature spectra had two bands with energies of about 1.1 and 1.2 eV at the maxima. The first maximum was attributed to $E_t = E_v + 0.33$–0.43 eV, and the second to $E_v + 0.21$ eV.

The results of an investigation of the optical absorption spectra of GaAs:Ni samples were reported in Refs. 220 and 249.

Absorption bands typical of the photoionization of a hole by a deep impurity centre were observed. The long-wavelength parts of the spectra indicated the presence of levels at 0.45 eV (80 K) and 0.9 eV (80 K), which were reported in Ref. 249. Since the electron–phonon interaction broadened the impurity absorption edge, it was suggested in Ref. 220 that the activation energy of the deep impurity should be deduced from the position of the impurity absorption band although it would be more correct to use Eq. (37). The activation energies obtained in this way were 0.42 eV (Ref. 220) and 0.35 eV (reported in a thesis* by Foigel').

We can see therefore that the scatter of the values of E_t attributed to nickel in gallium arsenide is very large.

The discrepancies between the reported values of E_t are primarily due to the absence of data on the concentrations of the dopant and of the residual donors in a given material. Moreover, in most studies only one particular investigation method has been used and this is clearly insufficient for obtaining reliable quantitative information on the energy spectrum of local levels in the system. Therefore, comprehensive investigations[71,253] employing simultaneously several methods (including ESR) are those worth the greatest attention in dealing with the nickel impurity atoms in the $Ni^{3+}(3d^7)$ state.

The samples used in these investigations were single crystals of GaAs:Ni prepared by the Czochralski method with liquid sealing of the melt by a layer of boron oxide. Nickel was introduced during growth. The dopant concentration was monitored by chemical analysis and it was found to be within the range $(3\text{-}15)\cdot 10^{16}$ cm^{-3}. Some of the crystals were deliberately doped with both nickel and tellurium, so that low-resistivity n-type samples were obtained. In the case of the weakly compensated crystals the total concentration of nickel was practically identical with the concentration of the Ni^{3+} ions. Consequently, all the nickel impurity centres were in a substitutional solid solution and they occupied the regular sites in the Ga sublattice. In the case of the crystals with the highest resistivity the degree of compensation was so high that practically all the nickel impurity centres were in the state with the $3d^8$ configuration, i.e. they were the Ni^{2+} ions.

Figure 74 shows the temperature dependences of the Hall coefficient of the investigated samples. In the case of the less

* M. G. Foigel', "Investigation of the absorption of light by deep impurity centres in semiconductors," Author's abstract of thesis for Candidate's Degree [in Russian], Odessa (1973).

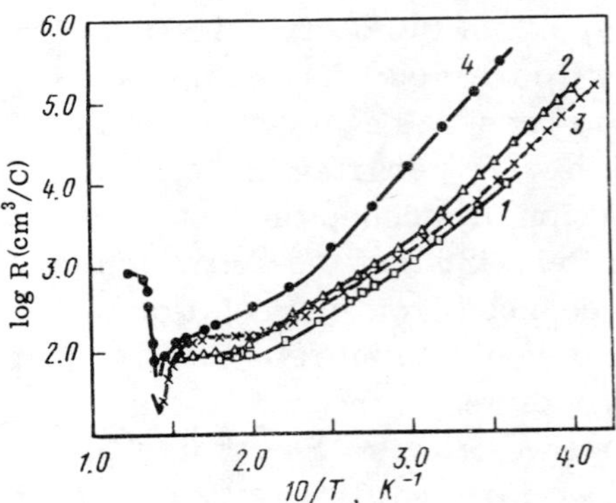

FIG. 74. Temperature dependences of the Hall coefficient of GaAs:Ni samples with different values of the concentrations of the Ni^{3+} ions (10^{16} cm^{-3}) and Ni^{2+} ions (10^{16} cm^{-3}) and of the electrical resistivity ρ (Ω.cm), respectively: 1) 5.0, 0.5, 15; 2) 5.0, 0.5, 24; 3) 4.0, 0.5, 13; 4) 2.9, 7.1, 460.

compensated samples with a relatively low conductivity the dependences R(1/T) has three characteristic regions with integral and half-integral slopes and with impurity exhaustion (curve 1). This made it possible to determine the characteristic parameters of deep centres: the ionization energy was E_t = 0.35 eV and the effective degeneracy factor was g^* = 10. The density of holes in the impurity exhaustion region p_{exh} agreed quite well with the Ni^{3+} concentration deduced from the ESR spectra. Hence, the values of E_i and g^* found in this way did indeed represent the nickel impurity levels.

In the case of the GaAs:Ni crystals with the higher resistivity the impurity exhaustion plateau was not exhibited by the R(1/T) curves and the slope of these curves was slightly higher. This was explained in Refs. 71 and 253 by the influence of deep states of residual (accidental) impurities of defects with energy levels located near E_v + 0.35 eV. In fact, as the nickel levels became filled with electrons (particularly when the nickel concentration in a given material was low), the role of these additional centres increased and under certain conditions they could govern the equilibrium properties of the crystals. For example, the levels at E_v + 0.44 eV or at E_v + 0.42 eV observed for GaAs:Ni (Table XXV) were in our opinion due to the presence of such additional centres not related to nickel.

The slopes of some of the curves in Fig. 74 amount to 0.38–0.39 eV. These values correspond to the half-sum of the slopes 0.35 and 0.44 eV which determines the temperature dependences of

TABLE XXVI. Energy Levels of Cobalt in Gallium Arsenide

$E_v + E_t$, eV	Method	Ref.	$E_v + E_t$, eV	Method	Ref.
0.16	H	[76]	0.56	H; OA; PH	[256]
0.16	H; PH	[251]	0.5	CL	[223]
0.16	OA	[255]	0.16*	H; OA; ESR; MS	[257]
0.54	TS	[254]			

Note. The notation is the same as in Tables XXIII-XXV, with the exception of MS which denotes the magnetic susceptibility.

the equilibrium parameters R and σ in the case of closely spaced levels.

The attribution of the $E_v + 0.35$ eV level to the Ni^{3+} ions[71,253] was confirmed also by investigations of the optical absorption and photoconductivity. However, the most convincing proof that the $E_v + 0.35$ eV level was due to the nickel impurity was provided by the observation that the intensity of the ESR signal corresponding to the Ni^{3+} ions was directly linked to the occupancy of the level, which was varied by the double doping method.

Cobalt

The values of E_t deduced for GaAs:Co samples by various authors are listed in Table XXVI. Haisty and Cronin[76] used the temperature dependence of the Hall effect to determine the activation energy of the cobalt impurity in GaAs and they obtained 0.16 eV. The same level was deduced in Ref. 251 from the temperature dependence of the Hall coefficient and from the photoconductivity spectrum. This was in agreement with the result obtained by Baranowski et al.[255] who estimated the activation energy of cobalt in GaAs from the position of the fundamental absorption edge. However, very different values of the activation energy of cobalt in GaAs were reported in Refs. 220, 223, and 254, where levels with activation energies 0.5-0.56 eV were reported. Unfortunately, the concentrations of cobalt in these samples were not given.

The results of a comprehensive investigation were reported in Ref. 257. The main distinction of this investigation is the reliability of the determination of the concentrations of cobalt in both $Co^{3+}(3d^6)$ and $Co^{2+}(3d^7)$ states. In the latter case the concentration was found by the ESR method, whereas in the former case it was deduced from the magnetic susceptibility. The overall concentration of cobalt atoms in GaAs was found by chemical analysis.

It therefore follows that the $E_v + 0.16$ eV level[257] with the degree of occupancy correlated with changes in the intensity in the

TABLE XXVII. Energy Levels of Manganese in Gallium Arsenide

$E_v + E_t$, eV	Method	Ref.	$E_v + E_t$, eV	Method	Ref.
0.094	H	[76]	0.11	OA	[258]
0.10*	H	[230]	0.07–0.11	H	[312]
0.091	H; PH	[251]	0.1–0.11	PL; CL	[312]
0.12	PL	[240]			

Note. The notation is the same as in Tables XXIII-XXV.

ESR spectrum and in the magnetic susceptibility is the most reliable value.

Manganese

Crystals of GaAs:Mn have not yet been investigated sufficiently thoroughly. According to the reported results, the manganese impurity in GaAs forms the shallowest acceptor level among the T-metal impurities (Table XXVII). The scatter of the values reported by different authors is slight and it is practically within the limits of the experimental error. Therefore, the most probable value of E_t is 0.1 eV.

Vanadium

Among all the transition metal impurities in GaAs the least known is vanadium. An investigation of samples prepared by the Czochralski method was reported in Ref. 259. The vanadium dopant was introduced during growth of the ingots; its concentration was deduced by chemical analysis and it amounted to $\sim(1\text{-}2)\cdot 10^{17}$ cm^{-3}. The starting material was n-type GaAs with an electron density $\sim 5\cdot 10^{16}$ cm^{-3}. Single crystals of GaAs:V had n-type conduction, had a high electrical resistivity of $\rho \sim 10^8$ Ω.cm, and the electron mobility in these crystals was $\sim 3\cdot 10^3$ $cm^2\cdot V^{-1}\cdot sec^{-1}$ at 300 K. Conversion of low-resistivity n-type GaAs into a semi-insulating material by introduction of vanadium demonstrated clearly that the dopant creates deep acceptors in the band gap, and not donor states as initially suggested in Ref. 108. The thermal ionization of the levels corresponding to the $V^{2+} \rightarrow V^{3+} + e$ transition was deduced in Ref. 251 from the temperature dependences of the equilibrium values of the parameters R and ρ in the range 300-800 K; its value was found to be $E_c - E_t = 0.77 \pm 0.03$ eV or $E_v + 0.87$ eV. It follows that the vanadium impurity (like chromium) exhibits the deepest levels located practically in the middle of the band gap of GaAs.

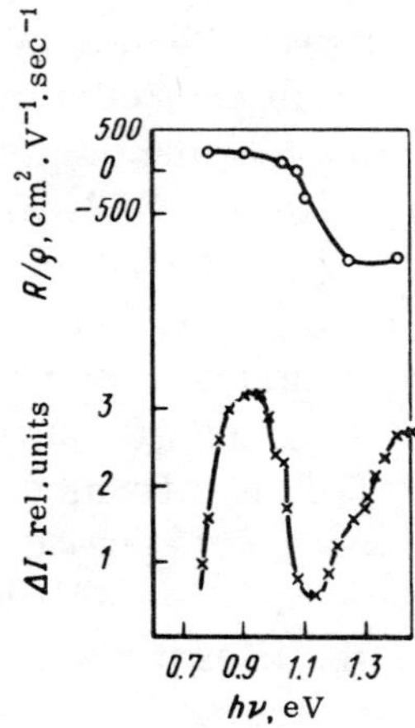

FIG. 75. Spectral dependences of the effective mobility of carriers and of the photoconductivity ΔI of GaAs:V samples at 160 K.

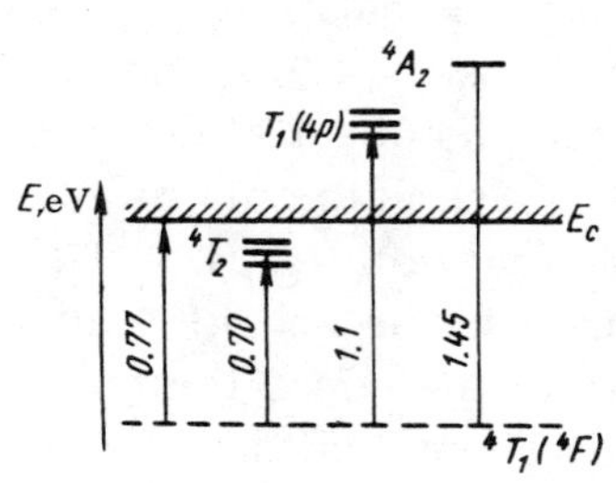

FIG. 76. Energy spectrum of local vanadium impurity states in GaAs.

The energy spectrum of the vanadium impurity in GaAs was also investigated using optical and photoelectric properties.[251] Figure 75 shows the spectra of the impurity conductivity and of the effective mobility R/ρ of nonequilibrium carriers. Near the long-wavelength edge at $h\nu < 1$ eV the photoconductivity of this material was due to nonequilibrium holes created as a result of optical transitions of electrons from the valence band to neutral $V^{3+}(3d^2)$ impurity centres. The optical activation energy deduced from the photosensitivity edge of the photoconductivity and absorption spectra was ~0.8 eV and it agreed quite well with the thermal ionization energy of deep vanadium levels. In the region of $0.8 < h\nu < 1.0$ eV the photoelectron density n was low compared with the density of nonequilibrium holes, $n \ll p$, because in the opposite case the much higher mobility of electrons in the range $n > 10^{-2}p$ would have given rise to n-type photoconductivity (as deduced from the sign of the Hall effect). The relationship between n and p found in this way can be explained by assuming that the photoionization cross-sections of an electron σ_n and a hole σ_p of the d-state of the vanadium impurity resulting in the transfer to the s-like bottom of the conduction band and to the p-like top of the valence band are related by the inequality $\sigma_n \ll \sigma_p$ which follows from the selection rules.

However, at high photon energies in the range $1 < h\nu < 1.3$ eV the photosensitivity fell significantly and the sign of the photoconductivity changed in the same spectral range: the majority carriers became electrons. Hence, the absorption band at 1-1.2 eV was due to intracentre optical transitions with excited states of the V^{2+} ion which were in resonance with the continuous spectrun of states in the conduction band. By analogy with the GaAs:Cr system, the intracentre absorption channel in the case of the vanadium doped samples led to a considerable increase in σ_n, i.e. the intensity of the photoexcitation of electrons to the conduction band increased. The autoionization effect released electrons from pseudolocal impurity states in ~10^{-13} sec and these free electrons could participate in the charge transport processes. The appearance of carriers of the opposite sign could alter the effective mobility and enhance recombination, i.e. it could reduce the lifetime and photosensitivity of the material, as indeed found experimentally.

The results obtained in Ref. 251 led the authors to propose the structure of vanadium levels in GaAs in the form shown in Fig. 76.

Scandium and Titanium

No data are available on the energy levels of these impurities in gallium arsenide.

Conclusions

A logical conclusion of this section would be a systematization of the results obtained similar to that represented by the dependence of E_t on the degree of occupancy of the d-shell of the T-metal ions in germanium and silicon (Fig. 68). However, a similar diagram plotted for GaAs:T-metals shown in Fig. 77 yields unexpectedly not one but two such dependences. It is worth noting that the majority of the ions exhibiting the Jahn–Teller effect lie on the rising curve. However, this observation still does not account for the very different nature of the variation of the energy spectrum of two groups of T-metal ions as their d-shell is filled. Moreover, it is not possible to explain either the behaviour of T-metal ions in GaAs and in elemental group IV semiconductors. In the latter the filling of the d-shell makes the $E_v + E_t$ level deeper (Fig. 68), whereas in the case of GaAs at least some of the ions exhibit the opposite tendency of $E_v + E_t$ (Fig. 77).

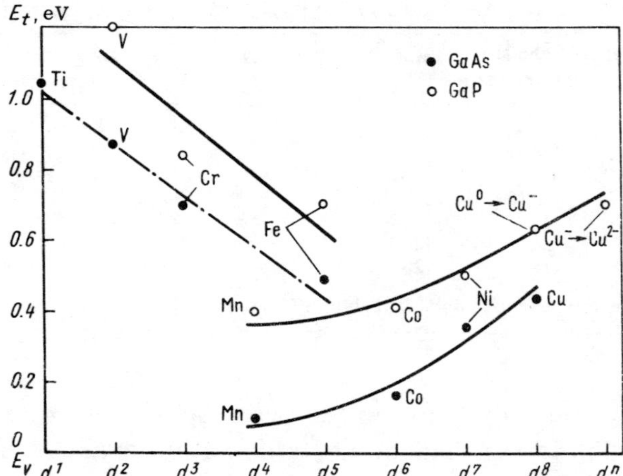

FIG. 77. Dependences of the ionization energies of T-metal impurities in GaAs and GaP on the degree of occupancy of the d-shell of the T-metal atoms.

3.4. ENERGY SPECTRUM OF LEVELS OF T-METAL IMPURITIES IN GALLIUM PHOSPHIDE

It is demonstrated in Sec. 1.5 that T-metal impurities in the crystal lattice of GaP behave exactly as in GaAs. Consequently, they form deep acceptor levels in the band gap of GaP. The energies of these levels are listed in Table XXVIII.

Iron

This is the most thoroughly investigated impurity in GaP. We shall consider the main results obtained for GaP:Fe on the basis of Ref. 78 because of the fullest utilization of the "arsenal" of the available investigation methods.

Figure 78 shows typical temperature dependences of the equilibrium parameters of GaP:Fe crystals grown by the Czochralski method and doped during growth. In the investigated range of temperatures (the upper limit of which was set by the evaporation of phosphorus from the samples) there was no impurity exhaustion region. However, the observed inversion of the type of conduction and compensation of the donor background in the original n-type GaP by introduction of iron indicated that the levels in GaP:Fe were of the acceptor nature. The thermal ionization energy of a deep level deduced from the slope of the temperature dependences was $E_v + 0.7$ eV. This value differed considerably from $E_v + 1.7$ eV given in Ref. 85.

TABLE XXVIII. Energy Levels of T-Metal Impurities in Gallium Phosphide

Impurity	E_v+E_t, eV	Method of determination	Ref.	Impurity	E_v+E_t, eV	Method of determination	Ref.
Fe	0.7*	C; H OA; ESR	[78]	Cr	0.84	C	[68]
	1.7–1.8	PH	[262]	Ni	0.5	OA	[86]
	1.7	PESR	[85]	Co	0.41	?	[261]
	0.3	C	[260]	Mn	0.4	C; H; OA	[84]
	1.2	OA	[260]	V	1.2	C	[83]
	0.84	OA	[263]				

Note. The notation is explained in Tables XXIII and XXIV.

The discrepancy between the two values was explained in Ref. 78 by postulating that the photoionization cross-section for illumination of photons of energy $h\nu \approx E_t$ was small and a change $\Delta N(Fe^{3+})$ was much smaller than the concentration itself $N(Fe^{3+})$, which – according to a spectrochemical analysis – was high in GaP (Ref. 78) amounting to $5 \cdot 10^{17}$–$2 \cdot 10^{18}$ cm^{-3}. A large change $\Delta N(Fe^{3+})$ (~10%), sufficient for detection by the ESR method, could be manifested only for $h\nu > E_t$.

However, this discrepancy could also be due to a strong overcompensation of the samples with specially introduced tellurium or tin donors, so that the samples had n-type conduction. These samples exhibited an additional B-type ESR spectrum (Sec. 1.5). The level found at E_v + 1.7 eV (Ref. 85) could be due to associated defects of the iron–donor type.

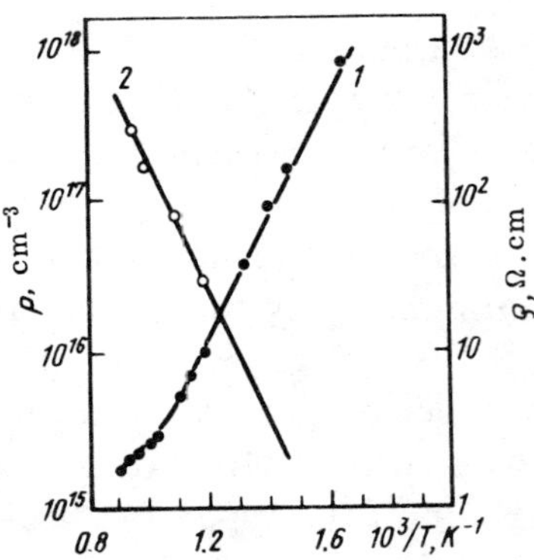

FIG. 78. Temperature dependences of the electrical resistivity (1) and of the hole density (2) in GaP:Fe samples.[78]

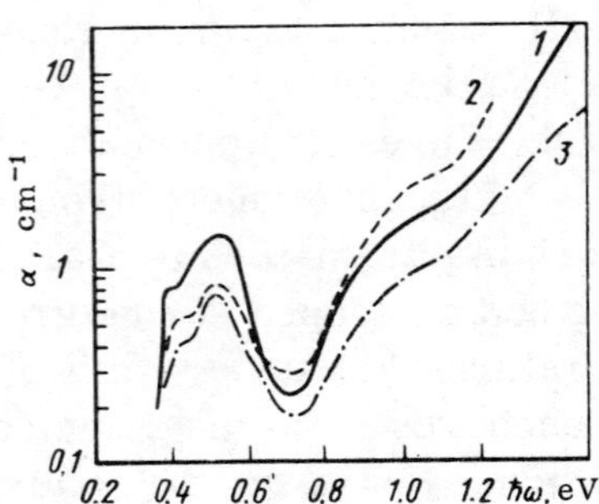

FIG. 79. Optical absorption spectra of GaP:Fe samples with different iron concentrations, recorded at 300 K (Ref. 78). The concentration of the Fe^{2+} ions rises with the curve number (1-3).

Figure 79, taken from Ref. 78, shows the absorption spectra of the same samples. These spectra were found to have three bands. The first – in the region of 0.4-0.6 eV – was of resonance nature typical of crystals containing iron in the $Fe^{2+}(3d^6)$ state. This band was associated with intracentre transitions and it was discussed in detail in Sec. 1.5. Its fine structure observed at low temperatures was also given earlier (Fig. 44).

The differences between the intensities of this band from sample to sample (Fig. 79) were well correlated with the donor background that governed the Fe^{2+} concentration. The amount of iron in the $Fe^{3+}(3d^5)$ state was deduced from the ESR signal intensity. The results obtained in this way agreed well with the optical absorption coefficient at $\hbar\omega$ = const corresponding to the charge transfer band (second band in the optical absorption spectrum in Fig. 79) at $h\nu > 0.7$ eV. Hence, the absorption of light was due to optical transfer of electrons from the valence band to the main iron level at $E_v + 0.7$ eV.

Finally, the third band in the optical absorption spectrum (Fig. 79) was interpreted in Ref. 260 as representing the optical ionization energy of an iron level at $E_t = 1.2$ eV, which differed greatly from the thermal ionization energy found by the same authors and amounting to 0.3 eV because of the electron–phonon interaction. This band was interpreted by Sobolevskiĭ in a different way* namely as a transition of an electron from the valence band to an excited level of the same iron centre. The absence of detailed investigations of the optical absorption in the $h\nu \geq 1.2$ eV range and of studies of the photoconductivity and photo-Hall effect in this range prevented reliable determination of the nature of the band.

Chromium

The temperature dependence of the electrical resistivity of GaP:Cr was found to vary exponentially with temperature and the activation energy was $E_v + 0.84$ eV (Ref. 68). The optical absorption spectra of GaP:Cr, shown in Fig. 80, were very similar to the spectrum of GaP:Fe. The first band in the region of 0.95-1.0 eV was also due to intracentre transitions occurring within the Cr^{2+} ions.

The second band clearly reflected the transfer of an electron from the valence band to the level E_t. However, it was very difficult to separate this band from the overall optical absorption

*V. K. Sobolevskiĭ, "Investigation of deep centres in gallium and indium phosphides," Author's abstract of thesis for Candidate's Degree [in Russian], Leningrad Polytechnic Institute (1979).

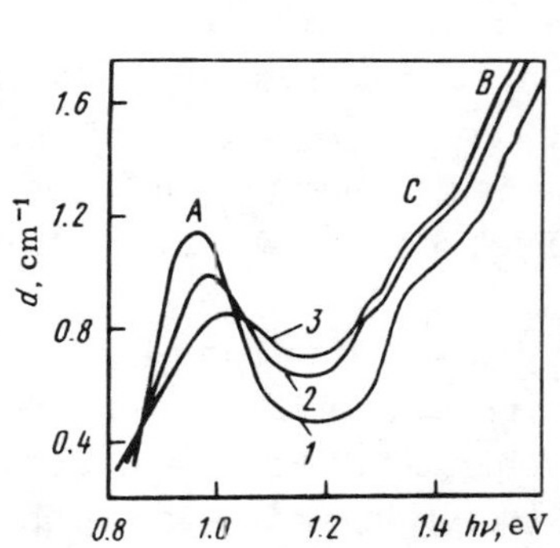

FIG. 80. Optical absorption spectra of GaP:Cr with a chromium concentration of $1.8 \cdot 10^{18}$ cm^{-3} recorded at temperatures 90 K (1), 300 K (2), and 400 K (3). Data taken from Ref. 68.

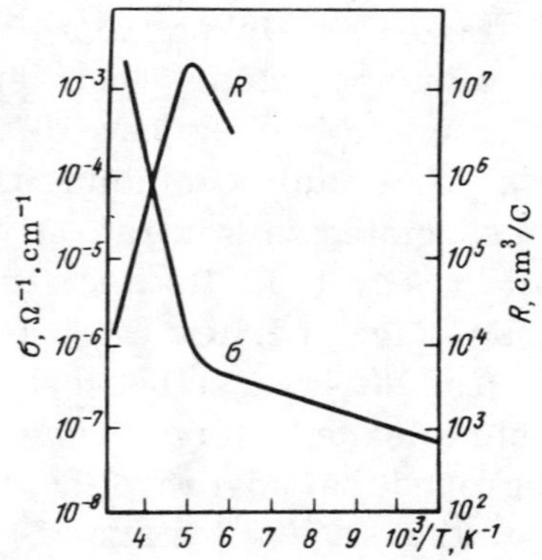

FIG. 81. Temperature dependences of the electrical conductivity σ and of the Hall coefficient R of GaP:Mn (Ref. 84). Manganese atom concentration $5.0 \cdot 10^{18}$ cm^{-3}.

spectrum and determine reliably the value of E_t.

The third band was due to an electron transition from the ground state of Cr^{3+} to the conduction band. The energy acquired for this optical ionization could, firstly, differ considerably from the thermal energy because of the electron–phonon interaction and, secondly, it was impossible to determine its value from the spectral curves in Fig. 80 without the penalty of a large error.

Nickel

This impurity has not yet been investigated sufficiently in GaP. Only the data on the optical absorption spectra of GaP:Ni samples are available.[86,226] In the first of these investigations we found the acceptor level of nickel and the ionization energy of the ground state was found to be $E_v + 0.5$ eV.

Cobalt

According to Ref. 261, cobalt forms an acceptor level at $E_v + 0.41$ eV in the band gap of GaP. Unfortunately, detailed investigations of the energy spectrum of levels in GaP:Co have not yet been reported.

Manganese

Both optical and electrical properties of GaP:Mn single crystals grown from the melt were considered in Ref. 84. The temperature dependences of the equilibrium parameters are plotted in Fig. 81.

They describe quite satisfactorily the ionization of one level $E_v + 0.4$ eV. A maximum on the dependence $R(10^3/T)$ reported in Ref. 84 is attributed to conduction involving deep impurity levels.

The value $E_v + 0.4$ eV is in practice reproduced also when the optical absorption edge is determined.[84] It is interesting to note that manganese in both GaP and GaAs creates the lowest level among all the T-metal adatoms.

The attribution of the $E_v + 0.4$ eV level to the Mn^{3+} ion or to the ionized (compensated) state Mn^{2+} is not yet fully established. An analogy of the behaviour of manganese in GaP and GaAs suggests equivalence of this state to the Mn^{3+} state and investigations of the optical absorption and ESR spectra in diffused GaP:Mn film provide evidence in preference of the $Mn^{2+}(3d^5)$ model. Clearly, the final answer can only be obtained by investigating photo-induced electron spin resonance.

Vanadium

Introduction of vanadium into gallium phosphide resulted in strong compensation of the material which had initially n-type conduction.[83] The room-temperature electrical resistivity of the samples reached 10^{12} Ω.cm. The thermal activation energy of electrical conduction was $E_v + 1.2$ eV. There was also a significant rise of the photoconductivity in the photon energy range 1.4-2.2 eV (Fig. 82) due to the transfer of electrons from the ground state of V^{2+} to the conduction band. Consequently, the vanadium impurity in GaP gave rise to the deepest level among all the T-metal impurities and this level was located practically in the middle of the band gap. It is possible that the photoconductivity edge exceeded by $E_c - 1.4$ eV the value of $E_g - (E_v + E_t)$ and this was due to the electron–

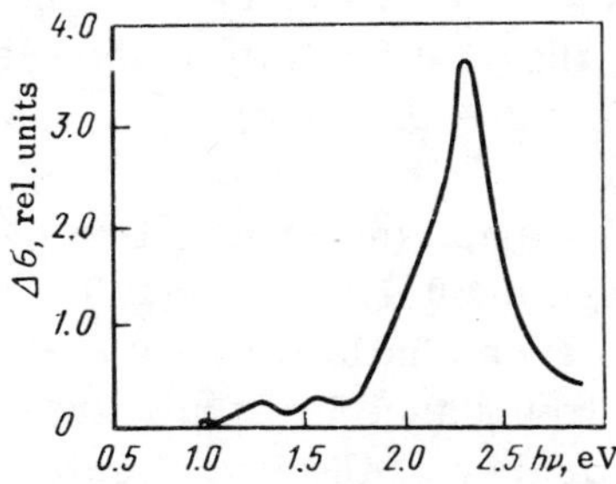

FIG. 82. Photoconductivity of GaP:V at 300 K (Ref. 83).

phonon interaction. However, the size of the excess representing a Franck–Condon shift of 0.32 eV seemed to be too large. The most reliable ionization energies of T-metal impurities in GaP are plotted in Fig. 77. Clearly, the dependences $E_t(d^n)$ are similar for GaP and GaAs. This reflects the similarity of the states of the corresponding T-metal atoms in these crystals.

3.5. ENERGY LEVELS OF T-METAL IMPURITIES IN INDIUM PHOSPHIDE

Practically all the investigations of InP:T-metal crystals have been carried out on samples characterized by strong compensation even in the absence of T-metal impurities. It was therefore stated in Ref. 177 that "for that degree of compensation which exists in InP crystals prepared so far, the opportunities for investigating deep impurities are very limited." This was true in the early seventies and it is largely true today.

For these reasons the values of E_t reported for several T-metal impurities in InP are very doubtful. We shall nevertheless review them briefly.

Iron

Investigations of electrical and photoelectric properties of InP:Fe have revealed acceptor levels (reduced by us to E_v): $E_v + 0.73$ eV (Ref. 264), $E_v + 0.56$ eV and $E_v + 0.83$ eV (Ref. 265). Thermal quenching of the photoconductivity of these samples was observed at $h\nu = 0.58$ eV, in agreement with the energy of the level at $E_v + 0.73$ eV. The investigation reported in Ref. 266 was concerned with Au–n-InP:Fe barriers which were optically excited to reveal levels at $h\nu_1 = 0.72$ eV, $h\nu_2 = 0.74$ eV, $h\nu_3 = 0.9$ eV, and $h\nu_4 = 0.54$ eV; thermal excitation showed that there were levels at $E_1 = 0.66$ eV, $E_2 = 0.71$ eV, $E_3 = 0.69$ eV, and $E_4 = 0.4$ eV.

The method of capacitance relaxation of similar barriers was used to find other levels: 0.79, 0.87, 1.04, and 1.14 eV (Ref. 267), whereas in Ref. 268 even shallower levels at 0.06 and 0.1 eV were reported. Such an "abundance" of the values of E_t is additional evidence of the low quality of the investigated samples.

The most probable value of E_t of the iron impurity in InP is in our opinion $E_v + 0.73$ eV (or $E_c - 0.58$ eV).

Chromium

The Hall effect measurements and a study of the photoconductivity made it possible to identify four levels[264]: $E_v + 1.24$ eV, $E_v + 0.79$ eV, $E_v + 0.69$ eV, and $E_v + 0.33$ eV. Only the second of these was attributed in Ref. 264 to the Cr^- ion. The last level was undoubtedly due to the Cu^- ion, because the value of E_t for this ion is well known.

The origin of the other two levels was not clear.

Nickel

Three levels were found in InP:Ni samples in Ref. 265 and their values reduced to the top of the valence band where $E_v + 0.35$ eV, $E_v + 0.57$ eV, and $E_v + 0.76$ eV. The first of these was attributed to nickel in Ref. 265.

Cobalt

The photoconductivity and infrared quenching (of the photoconductivity) spectra were investigated.[269] The photoconductivity maximum was observed at $h\nu \geq 1.2$ eV and infrared quenching at $h\nu \leq 1.2$ eV. On this basis the level at $E_c - 1.2 = E_v + 0.21$ eV was attributed to cobalt in InP.

Manganese

An investigation of the characteristics of S-type diodes made of InP:Mn allowed us to draw the conclusion[269] that manganese was an acceptor with an activation energy $E_v + 0.4$ eV and the value of $E_v + 0.26$ eV deduced from the optical absorption was reported in Ref. 88; the latter corresponded to the $Mn^{2+}(3d^5)$ state. This seemed to us to be more reliable because it was obtained for single crystals rather than for p-n junctions.

The following values of E_t are in the opinion of the investigators due to T-metal ions in InP:

Impurity	Fe	Cr	Ni	Co	Mn	Mn
$E_v + E_t$, eV	0.73	0.79	0.35	0.21	0.4	0.26*
Reference	265	264	265	265	269	88

The insufficiency of the data and scarcity of information on the state of T-metal ions in InP (Sec. 1.6) make it impossible to judge the relationship between E_t and the degree of occupancy of the d-shells of these ions in InP.

3.6. ENERGY SPECTRUM OF LEVELS OF T-METAL IMPURITIES IN NARROW-GAP III-V SEMICONDUCTORS

In Sec. 1.6 we have already mentioned that the behaviour of T-metal impurities in narrow-gap III-V compounds (InAs and InSb) is quite different from the behaviour in wide-gap compounds GaAs, GaP, and InP. Impurities in the form of T-metals (at least iron and chromium) exhibit donor and not acceptor properties in narrow-gap III-V semiconductors.[91] It should be pointed out that this conclusion is disputed in Ref. 92. It follows that the present status of our knowledge of the states of iron, chromium, and other T-metal impurities in InAs and InSb is still subject to doubt.

Nevertheless, serious attempts have been made to determine E_t of these impurities from measurements of equilibrium parameters. For example, an investigation of the galvanomagnetic properties of InSb doped with manganese and iron indicated that these impurities form shallow acceptor levels with E_t amounting to $E_v + 0.009$ eV and $E_v + 0.016$ eV, respectively. However, in all the investigations of InSb:T-metal the samples had n-type conduction and a low resistivity. As soon as high-resistivity n-type samples were prepared by introduction of a shallow acceptor simultaneously with a T-metal impurity,[91] it became possible to determine the parameters of deep T-metal levels. However, a reliable determination was difficult because of the shunting action of the surface conductance of these crystals.

Measurements reported in Ref. 91 were carried out on samples of n-type InSb with electron densities $n < 10^{14}$ cm^{-3} at 77 K. The high electrical resistivity and a residual donor background of $\sim 1 \cdot 10^{14}$ cm^{-3} were achieved by simultaneous introduction of chromium and zinc impurities or of chromium and manganese, which – like zinc – gave rise to shallow acceptor states in InSb. By comparison, measurements were also made on high-resistivity crystals strongly compensated with the germanium impurity (shallow acceptor). Before measurements the samples were etched in the traditional CP-4 solution which reduced greatly the resistance. After etching the resistivity ρ continued to rise with time and reached its steady-state value after approximately 24 hours. Nevertheless, the surface leakage on these crystals remained significant and it gave rise to a number of anomalies considered and explained quantitatively in Ref. 270. An important experimental observation was the weak influence of surface leakage for some time immediately after etching of the samples in Trilon B (ethylenediamine of tetraacetic acid). This made it possible to determine the temperature dependences of the electron density (Fig. 83) and to estimate the position of the chromium level in InSb, which was $E_c - 0.074$ eV.

Figure 84 shows the steady-state photoconductivity spectra of InSb:Cr obtained at 80 K. Clearly, the long-wavelength edge of the impurity photoconductivity agreed well with the thermal energy of ionization of the deep chromium level.

In the case of the crystals with the highest resistivity and with $n < 10^{12}$ cm^{-3} there was a strong negative photoconductivity band at 0.13-0.19 eV. Its appearance was due to the fact that the double optical transitions of both electrons and holes from the chromium levels to the allowed bands occurred simultaneously in this part of the spectrum. Consequently, the photoconductivity became strongly ambipolar and the electron lifetime was no longer governed by the recapture by the E_t level, but by the capture by other recombination centres responsible for nonequilibrium processes in InSb. It should be pointed out that the negative photoconductivity band began at photon energies somewhat lower than the band gap between E_t and the top of the valence band, i.e. several phonons should participate in these transitions. It would seem that the negative photoconductivity could be manifested only at $h\nu \gtrsim E_v + E_t =$

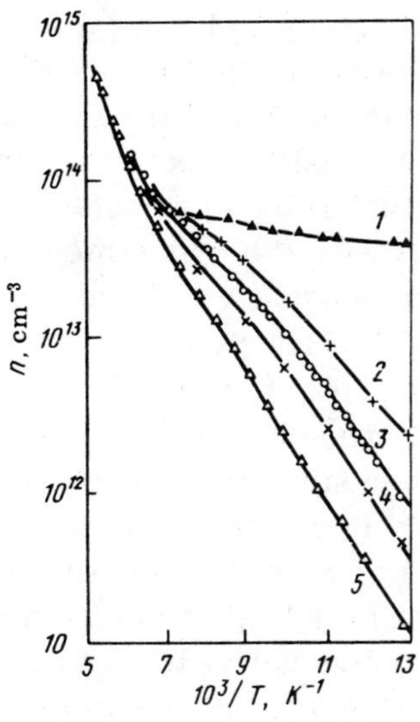

FIG. 83. Temperature dependences of the electron density in InSb:Cr samples. The curves are labelled in accordance with the different values of the resistivity ρ (77 K), in Ω.cm: 1) 0.33; 2) 6.5; 3) 15.8; 4) 34.6; 5) 117.5. Dopants: 1) Cr; 2)-4) Cr and Zn; 5) Cr and Mn.

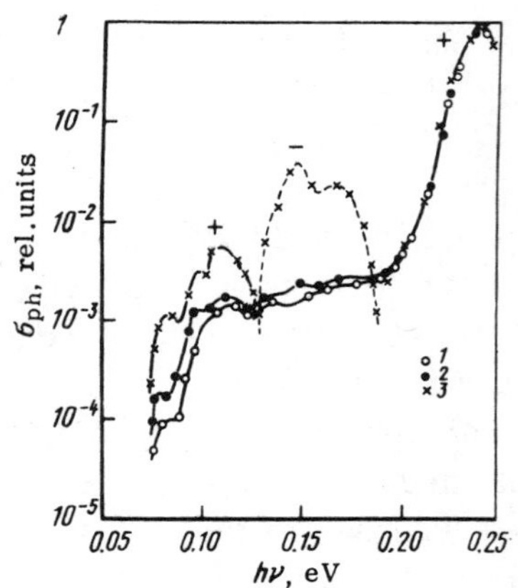

FIG. 84. Impurity photoconductivity of high-resistivity InSb samples at 77 K. Dopants and electron densities (cm^{-3}): 1) Cr, $1.2\cdot10^{13}$; 2) Cr and Mn, $2.1\cdot10^{13}$; 3) Cr and Mn, $1.4\cdot10^{11}$.

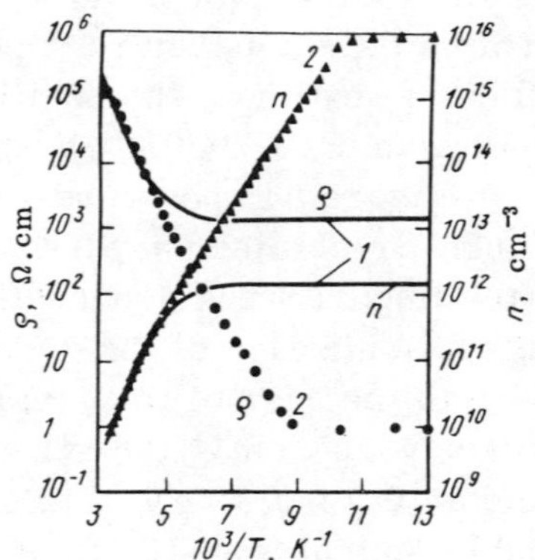

FIG. 85. Temperature dependences of the electrical resistivity and of the carrier density in n-type InAs:Cr:Zn before (1) and after (2) illumination with light corresponding to the fundamental absorption region.

0.15 eV. However, if we bear in mind that the photoionization cross-sections of electrons σ_n and holes σ_p representing transitions from the d-level of chromium to the s-like bottom of the conduction band and to the p-like top of the valence band are related by $\sigma_n \ll \sigma_p$, the proposed explanation of the shift of the negative photoconductivity band seems to be fully justified.

Within the framework of this model of negative photoconductivity it is natural to assume that filling of the chromium levels, i.e. an increase in the density of equilibrium electrons, should reduce the rate of optical release of holes from these levels. This was in good agreement with the experimental results because the negative impurity photoconductivity band was not exhibited by samples with $n > 10^{13}$ cm^{-3} (Fig. 84).

In the case of InAs:Cr the effects of surface leakage at low temperatures were suppressed by illumination of samples with strong light from the fundamental absorption region.[271] It was found that this resulted in a considerable reduction of the band bending on the surface of a crystal and of the surface charge density. It was important to note that the illumination-modified ratio of the bulk and surface conductivities was retained for at least 24 hours. The initial parameters of the samples were restored only after heating.

After preliminary illumination the temperature dependences $\rho(T)$ and $n(T)$ revealed an impurity conduction region (Fig. 85),

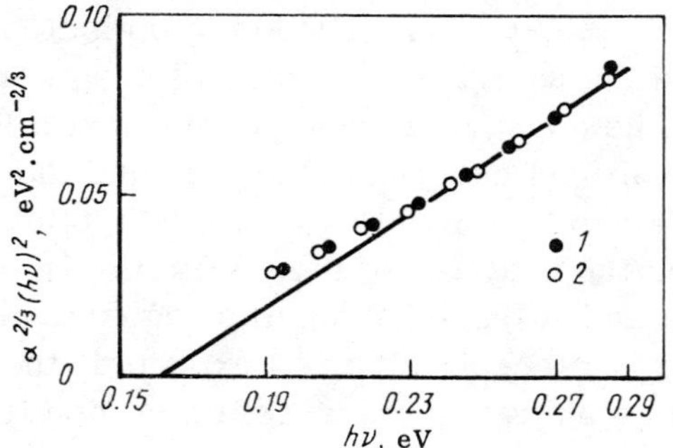

FIG. 86. Optical absorption spectrum of a low-resistivity n-type InAs:Cr sample recorded at 77 K (1) and 300 K (2). The continuous line represents a calculation based on Eq. (37).

which was used to determine[271] the thermal ionization energy $E_c - E_t = 0.15$ eV of the chromium level in InAs.

This value was also in agreement with an investigation of the optical absorption reported in the same paper[271] and represented in Fig. 86 by the same coordinate as in Eq. (37).

The discrepancy between the calculations and experiments observed in the region of absorption by free carriers at photon energies $h\nu < 0.22$ eV was clearly due to the different densities of electrons in the investigated crystal and in the sample placed in the comparison channel of the spectrometer.

Figure 87 shows the absorption spectrum of high-resistivity InAs:Cr:Zn crystals. The long-wavelength edge of the spectrum

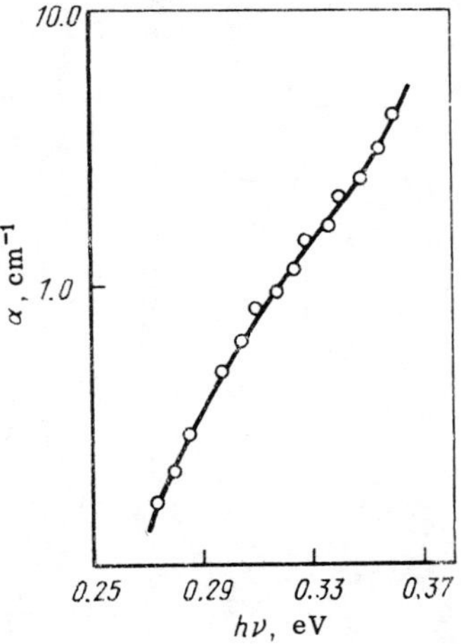

FIG. 87. Optical absorption spectrum of a high-resistivity n-type InAs:Cr:Zn sample recorded at 77 K.

was 0.26 eV and it clearly corresponded to optical transitions from the valence band to the partly filled states of deep chromium levels. The transitions to the conduction band occurred at a low rate compared with the transitions between chromium and the valence band. This could be explained by the selection rules applicable to optical transitions between the d-states of the chromium impurity and the allowed conduction and valence bands, the edges of which had the s- and p-symmetry, respectively. As expected, the sum of the two optical ionization energies was practically identical with the band gap of InAs.

The absence of data on other InSb:T-metal and InAs:T-metal systems prevents us from drawing any general conclusions on the energy spectra of transition-metal impurities in narrow-gap III-V semiconductors.

4. MAGNETIC EFFECTS IN SEMICONDUCTORS DOPED WITH TRANSITION-METAL IMPURITIES

4.1. MAGNETISM OF DOPED SEMICONDUCTORS

Elemental group IV semiconductors and III–V semiconductor compounds considered in the present monograph are weakly magnetic substances and according to the theory of magnetism their magnetic susceptibility χ can be represented by a sum of several paramagnetic and diamagnetic components. These components include the lattice susceptibility χ_{latt}, the susceptibilities of carriers χ_n and χ_p, and the susceptibility due to various defects (impurity centres, thermal defects, dislocations, etc.). The main task of an analysis of the magnetic properties of semiconductors is to separate these components.

In the case of metals such separation is very difficult because the strong degeneracy of the electron gas and the constancy of the electron density make the magnetic susceptibility (including the lattice susceptibility) independent of temperature.

In the case of semiconductors the task is greatly simplified mainly because it is possible to vary carrier density by doping a crystal.

For these reasons the values of χ_{latt} have now been determined for the majority of elemental semiconductors and III–V compounds. The lattice susceptibility is usually taken to be the susceptibility of highly purified and relatively "perfect" samples with impurity and defect concentrations not exceeding 10^{15} cm^{-3}.

The situation is more satisfactory in respect of the theory of the lattice magnetic susceptibility. According to the current ideas, a system of interacting atoms is regarded as composed of two subsystems: ionic cores forming the crystal lattice and a system of collective-state (itinerant) electrons which in nonmagnetic substances are the chemical-bond electrons.

In quantum mechanics the susceptibility of atoms and molecules without an intrinsic magnetic moment is described by the expression

$$\chi = -Ne^2/6m_0c^2 \sum_n (n|r_i^2|n) + 2N \sum_{n' = n} (n|\hat{M}_z^0|n')^2/(E_n^0 - E_{n'}^0), \quad (46)$$

where n and n′ represent the various states of the system; $(n|\hat{M}_z^0|n')$ is a matrix element of the magnetic moment operator taken for unperturbed states (in H = 0); $E_n^0 - E_{n'}^0$ are the energies of unperturbed states.

The first term in Eq. (46) describes the diamagnetic contribution of electrons (Langevin diamagnetism). The second describes the polarization paramagnetism associated with the virtual quantum transitions between zero (in H = 0) stationary states (Van Vleck paramagnetism). The diamagnetic term is a sum of the diamagnetic currents due to the individual electrons. Therefore, all the changes in this component of the susceptibility as a result of condensation of atoms to form a solid are due to the deformation of the electron orbits, i.e. they are due to changes in r_i. Such deformation decreases on approach of an electron orbit towards a nucleus. Therefore, the diamagnetism of filled electron shells is not greatly affected by the condensation of atoms into a solid.

The Van Vleck paramagnetism of ionic cores with a spherical symmetry is very weak, since the difference between the energies of the ground and excited states of electrons is usually large.

When atoms condense to form a solid, the changes should be much greater in the case of the valence electrons because the formation of chemical bonds distorts greatly the atomic orbits.

In nonmetallic crystals the chemical binding can be represented as hybridization of two limiting cases of ionic and covalent binding.

If the binding is purely ionic, we can assume that ions have the central symmetry and are not deformed. In this case the magnetic susceptibility of an ionic crystal should obey the rule of additivity of the susceptibilities of ions composing a crystal. In real crystals the distribution of the electron density is not spherical because of the electrostatic interaction between electron shells. This gives rise to the Van Vleck paramagnetic contribution. The size of this contribution increases on increase in the mutual deformation of electron shells, i.e. on increase in the departure from the purely ionic binding. The Van Vleck paramagnetism plays an even more important role in crystals with the covalent binding because of the anisotropy of the distribution of the valence electrons forming covalent "bridging" bonds.

A theory capable of providing a rigorous quantitative estimate of the lattice magnetic susceptibility is not yet available. Therefore, several semi-empirical methods have been suggested for the separation of the diamagnetic and paramagnetic contributions.[272-274]

In the Kirkwood–Dorfman method[272] use is made of the relationship between the diamagnetic component χ_d and the experimentally determined static electrical polarizability of crystals α:

$$\chi_d = -(Ne^2/4m_0c^2)\sqrt{a_0}\sqrt{Z\alpha}, \qquad (47)$$

where a_0 is the Bohr radius; Z is the number of electrons per atom; N is the concentration of atoms in 1 cm^3 of a crystal.

Kirkwood obtained this relationship for the specific case of centrally symmetric systems (atoms and ions), whereas Dorfman demonstrated that it was applicable also to polyatomic systems.

The paramagnetic contribution is found from the difference $\chi_p = \chi_{latt} - \chi_d$.

The Sirota method[273] is based on the X-ray determination of the distribution of the electron density in a crystal. Having determined the intensities of the Bragg reflections, we can find the values of the structure amplitudes and after summation of three-dimensional Fourier series, we can determine the spatial distribution of the electron density $\rho(x, y, z)$. Next, the electron density patterns plotted in this way can be used to calculate χ_d from the formula[273]

$$\chi_d = -35.565 \cdot 10^{10}\int_0^\infty \rho(r)r^4 dr, \qquad (48)$$

where $\mathbf{r}^2 = x^2 + y^2 + z^2$.

The advantage of the Sirota method is the ability to use the same electron density patterns to determine simultaneously and independently also the paramagnetic component χ_p:

$$\chi_p = (8/3)N\mu_B^2\delta/\Delta E, \qquad (49)$$

where δ is the index of nonsphericity of the distribution of the electron density; μ_B is the Bohr magneton. The quantity $\Delta E = E_{n'} - E_n$ for semiconductor crystals can be assumed to be equal to the band gap.

The difficulties encountered in the experimental determinations of the intensities of Bragg reflections and summation of the three-dimensional Fourier series reduce the accuracy of the Sirota method.

In the third semi-empirical method proposed in Ref. 274 the average internal potential of the lattice V_0 is deduced from the Kikuchi lines which appear in electron diffraction patterns when fast electrons are reflected from the surface of a crystal.

The relationship between χ_d and V_0 for a spherically symmetric (relative to a nucleus) distribution of electrons is[274]:

$$\chi_d = eMV_0/(4\pi m_0 c^2 \rho), \quad (50)$$

where M is the molecular mass of a crystal; ρ is the density of a crystal.

In addition to the lattice component, the total magnetic susceptibility may include a large contribution of free carriers. The magnetic susceptibility of free carriers also has both paramagnetic and diamagnetic components. Semiconductors may have parabolic (germanium, silicon) and nonparabolic (InSb, GaAs, etc.) dispersion laws governing the energy band structure and the electron gas may be nondegenerate or degenerate.

A theory of the magnetic susceptibility of the electron gas in semiconductors was developed furthest in Ref. 275. The paramagnetic component due to the spin motion has the following form in the case of an arbitrary spherical band and an arbitrary degree of degeneracy:

$$\chi_p = (n/4)(\mu_B^2/kT)\langle g^2 d/d\varepsilon\rangle, \quad (51)$$

where n is the electron density; g is the spectroscopic splitting factor of an electron in the conduction band; ε is the reduced energy of an electron equal to E/kT. The angular brackets in Eq. (51) denote integrals of the type

$$\langle A\rangle = \int_0^\infty (-\partial f_0/\partial E) A(E) k^3(E) dE, \quad (52)$$

where f_0 is the Fermi–Dirac distribution function and k is the wave vector.

The diamagnetic susceptibility of conduction electrons can be obtained from the Landau–Peierls formula:

$$\chi_d = (4\pi/3)(\mu_B^2 m_0^2/h^4)\int (\partial f_0/\partial E)[(\partial^2 E/\partial k_x^2)(\partial^2 E/\partial k_y^2) - (\partial^2 E/\partial k_x \partial k_y)^2]dk. \quad (53)$$

After integration over a solid angle in the k space, we obtain

$$\chi_d = -(8/3)(\mu_B^2 m_0^2/h^2)\langle (1/k^2)[(1/m_d) - (1/2m^*)]\rangle, \quad (54)$$

where m_d is the "dynamic" effective mass defined by

$$1/m_d = (1/\hbar^2)[(1/3)(d^2E/dk^2) + (2/3)(1/k)(dE/dk)]. \quad (55)$$

The ratio m^*/m_d can be represented in the form[276]

$$m^*/m_d = 1 - (1/3)(k/m^*)(dm^*/dk) = 1 - \gamma, \quad (56)$$

where the parameter γ is used as a measure of the deviation of the band from parabolicity.

Equation (54) can rewritten in the form

$$\chi_d = -(4/3)(\mu_B^2 m_0^2/h^2)\langle(1/k^2 m^*)(1 - 2\gamma)\rangle. \quad (57)$$

In the case of a III–V semiconductor, the conduction band of which is described by the Kane model, we have

$$K = (\sqrt{2m_n}/\hbar)E^{1/2}(1 + E/E_g)^{1/2}, \quad (58)$$

where m_n is the effective mass of an electron at the bottom of the conduction band. Then, introducing $E/E_0 = b$, we obtain

$$m^* = m_n(1 + 2b); \quad m_d = m_n(1 + 2b)^3/[1 + (8/3)b(1 + b)]; \quad (59)$$

$$\gamma = (4/3)[b(1 + b)]/[(1 + 2b)^2]. \quad (60)$$

For an arbitrary degree of degeneracy of the electron gas, we find that Eqs. (57), (59), and (60) yield the Zawadzki formula[275]:

$$\chi_d = -(8/3)(\mu_B^2 m_0^2/h^2 m_n)(2m_n kT/\hbar^2)^{1/2}[{}^0L_{-3}^{1/2} + 8/3\beta\,{}^0L_{-3}^{3/2} - 1/2\,{}^0L_{-1}^{1/2}], \quad (61)$$

where $\beta = kT/E_g$; ${}^nL_k^m$ represents generalized Fermi integrals given by

$${}^nL_k^m = \int_0^\infty(-\partial f/\partial Z)Z^n(Z + \beta Z^2)^m(1 + 2\beta Z)^k dZ. \quad (62)$$

The Zawadzki formula was simplified by Savel'ev[277] by simple algebraic transformations:

$$\chi_d = C[(1/2)\,{}^0L_{-3}^{1/2} + (2/3)\beta\,{}^0L_{-3}^{3/2}], \quad (63)$$

where C is of the same form as the complex coefficient in front of the brackets in (61). The integral ${}^0L_{-3}^{3/2}$ is tabulated in Ref. 278 and the integral ${}^0L_{-3}^{1/2}$ – in Ref. 277.

Less detailed results of calculations of these two integrals are given in Table XXIX. The expressions for a parabolic band can be obtained by substituting $\gamma = 0$ in Eq. (56) and the expression for χ_d in the case of strong degeneracy is derived from **Eq.** (57) by the substitution $k = (3\pi^2 n)^{1/3}$:

$$\chi_d = -(4/3)(\mu_B^2 m_0^2/h^2)[(3\pi n)^{1/3}/m^*](1 - 2\gamma). \quad (64)$$

In this case we have $\gamma = (n/m^*)(dm^*/dn)$.

The validity of Eq. (63) is illustrated well in Fig. 88 from Ref. 277.

When the conduction band is even more complex, as is true

TABLE XXIX. Values of Generalized Fermi Integral ${}^{0}L_{-3}$

η	*L* for values of β							
	0.00	0.01	0.02	0.03	0.05	0.07	0.09	0.12
				Values of ${}^{0}L_{-3}^{1/2}$				
−4.0	0.0160	0.0148	0.0138	0.0129	0.0115	0.01028	0.0093	0.0082
−3.0	0.0426	0.0394	0.0366	0.0342	0.0308	0.0272	0.0247	0.0217
−2.0	0.1096	0.1010	0.0937	0.0874	0.0772	0.0692	0.0626	0.0548
−1.0	0.2606	0.2388	0.2205	0.2049	0.1795	0.1597	0.1437	0.1248
0.0	0.5361	0.4855	0.4437	0.4084	0.3518	0.3084	0.2740	0.2339
1.0	0.9102	0.8073	0.7241	0.6564	0.5482	0.4685	0.4070	0.3375
1.2	0.9893	0.8728	0.7791	0.7021	0.5829	0.4949	0.4276	0.3520
1.4	1.0681	0.9369	0.8321	0.7464	0.6148	0.5185	0.4453	0.3637
1.6	1.1460	0.9991	0.8826	0.7879	0.6435	0.5389	0.4600	0.3727
1.8	1.2227	1.0591	0.9903	0.8263	0.6689	0.5560	0.4714	0.3789
2.0	1.2977	1.1166	0.9750	0.8614	0.6909	0.5698	0.4800	0.3824
3.0	1.6426	1.3611	1.1502	0.9817	0.7534	0.5961	0.4845	0.3689
5.0	2.1916	1.6652	1.3080	1.0540	0.7247	0.5270	0.3992	0.2786
7.0	2.6208	1.8170	1.3266	1.0063	0.6291	0.4255	0.3042	0.1982
10.0	3.1486	1.9051	1.2573	0.8816	0.4908	0.3058	0.2056	0.1250
15.0	3.8658	1.8856	1.0791	0.6834	0.3320	0.1895	0.1198	0.0684
20.0	4.4675	1.7838	0.9093	0.5339	0.2364	0.1278	0.0780	0.0430
				Values of ${}^{0}L_{-3}^{3/2}$				
−4.0	0.0242	0.0217	0.0198	0.0181	0.0155	0.0135	0.0120	0.0102
−3.0	0.066	0.0595	0.0540	0.0490	0.0423	0.0369	0.0327	0.0273
−2.0	0.1719	0.1542	0.1399	0.1280	0.1093	0.0952	0.0841	0.0714
−1.0	0.4357	0.3895	0.3522	0.3213	0.2729	0.2366	0.2083	0.1760
0.0	1.0171	0.9020	0.8098	0.7341	0.6167	0.5296	0.4625	0.3866
1.0	2.0946	1.8310	1.6232	1.4550	1.1988	1.0130	0.8724	0.7163
1.2	2.3795	2.0723	1.8313	1.6367	1.3419	1.1294	0.9693	0.7923
1.4	2.6881	2.3318	2.0534	1.8297	1.4924	1.2506	1.0694	0.8702
1.6	3.0202	2.6087	2.2889	2.0329	1.6491	1.3758	1.1720	0.9491
1.8	3.3756	2.9026	2.5369	2.2455	1.8111	1.5039	1.2761	1.0284
2.0	3.7537	3.2124	2.7963	2.4664	1.9775	1.6340	1.3809	1.1073
3.0	5.9655	4.9706	4.2301	3.6591	2.8409	2.2879	1.8930	1.4795
5.0	11.756	9.1967	7.4358	6.1640	4.4737	3.4235	2.7209	2.0298
7.0	18.997	13.875	10.647	8.4682	5.7827	4.2379	3.2627	2.3512
10.0	32.017	21.017	15.028	11.324	7.1794	5.0173	3.7378	2.6054
15.0	58.414	32.526	20.925	14.715	8.5700	5.7036	4.1200	2.7905
20.0	89.719	42.761	25.334	16.955	9.3502	6.0514	4.3004	2.8716

of InSb, it is necessary to allow for the merging of the higher bands. A theoretical analysis and the relevant expression valid in this case can be found in Ref. 275.

In the case of p-type semiconductors the high values of the effective masses make the magnetic susceptibility of holes small so that it is usually ignored compared with the other components of χ.

It should also be pointed out that a theoretical interpretation of the magnetic susceptibility of holes is difficult because the valence band has a more complex structure than the conduction band. As pointed out above, one of the components of the magnetic sus-

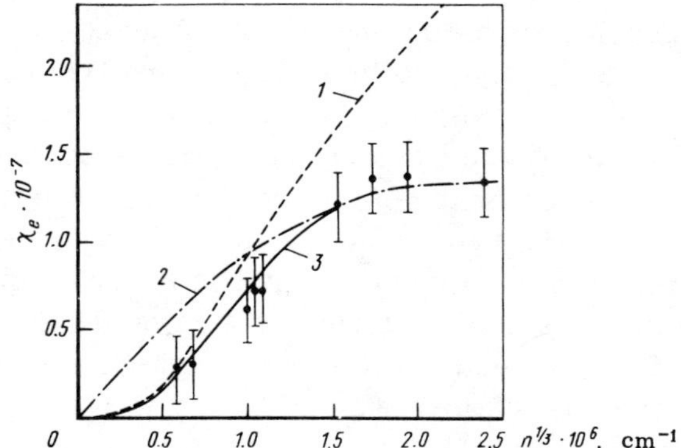

FIG. 88. Dependences of volume magnetic susceptibility of conduction electrons in gallium arsenide on the density of these electrons (T = 300 K), calculated: 1) for a parabolic band; 2) allowing for the band nonparabolicity in the heavy doping approximation; 3) from Eq. (78).

ceptibility is due to impurity atoms, which can make a considerable contribution only if the number of electrons in the outer shell differs from the number of electrons necessary for the formation of the valence bonds or if an impurity atom has partly filled inner shells.

The simplest example is the magnetism of noninteracting shallow donors and acceptors which are in the nonionized state at low temperatures. An impurity atom is then in the S-state (orbital momentum vanishes) and its magnetic properties are governed only by the spin momentum of the excess electron. The paramagnetic susceptibility is then described by the Langevin formula

$$\chi_p = N\mu_B^2/kT, \tag{65}$$

where N is the concentration of donors or acceptors.

Moreover, an excess electron exhibits the Larmor diamagnetism

$$\chi_d = -e^2N\overline{r^2}/6m^*c^2, \tag{66}$$

where $\overline{r^2}$ represents the average value of the square of the distance of the electron from the z axis directed along the magnetic field when the origin of the coordinate system is located at the atomic nucleus. It is clear from Eqs. (65) and (66) that at low temperatures the total diamagnetic susceptibility of semiconductors should decrease in accordance with the Curie law, which is indeed found experimentally in the absence of the impurity–impurity interaction.

A theoretical analysis of the magnetic susceptibility of interacting shallow donors considered in the approximation of an impurity band with a free electron Fermi gas can be found in Ref. 279. More realistic is the approximation in which the impurity interaction is considered as the formation of quasi-hydrogen molecules.[280] These molecules may be in a singlet or a triplet state (with an antiparallel or parallel orientation of the electron spins).

The difference between the energies of these states $\Delta\varepsilon$ is governed by the distance between the interacting donors in such a "molecule." The value of $\Delta\varepsilon$ is estimated in Ref. 280 as $\Delta\varepsilon$ for the hydrogen molecule with corrections for the permittivity and for the effective mass of electrons in a crystal.

It is thus found that the paramagnetism of interacting donors is described by the formula

$$\chi_p = C/T^{1-\alpha}, \tag{67}$$

where

$$C = (N\mu_B^2/k)(A/m^*)^{\alpha}(1 + \alpha)^{-1}; \quad \alpha = N/Bm^{*3}; \tag{68}$$

A and B are constants.

After the appearance of the work of Éfros and Shklovskiĭ[281] who applied the theory of percolation to the formation of an impurity band it has become necessary to rethink also the magnetism of interacting donors, especially as the errors in the determination of small values of χ_p make it impossible to detect the difference between $1/T$ and $1/T^{1-\alpha}$. The reduction in the Curie constant C cannot be treated unambiguously because of the indeterminacy of the g-factor. However, the general features of Eq. (67) are in good agreement with the experimental results.[280] The ground state of a partly filled d-shell in a transition metal of the iron group is governed by the Hund rules (Sec. 1.3). According to the first rule, an atom or ion with a given electron configuration has the largest total spin S and the largest (for given S) net orbital momentum L. The quantum number of the total momentum J is, in accordance with the second Hund rule, equal to L – S for a shell which is less than half-filled and L + S for a shell which is at least half-filled. The magnetic moment of a T-metal atom (ion) corresponding to given values of L, S, and J is defined by

$$\mu = g\sqrt{J(J + 1)}\mu_B, \tag{69}$$

where

$$g = 1 + [J(J + 1) + S(S + 1) - L(L + 1)]/[2J(J + 1)]. \tag{70}$$

The susceptibility of a paramagnetic T-metal impurity then obeys the Curie law:

$$\chi = N\mu^2/3kT. \tag{71}$$

The deviation from the Curie law may be due to the appearance of an additional paramagnetism as a result of the influence of the excited levels of a T-metal ion. In general, the paramagnetism is described by a term similar to the second term in Eq. (46). This contribution is independent of temperature as long as there is no change in the population of the levels.

Ions of T-metals in semiconductors may have different magnetic moments, depending on their position in the crystal lattice and on the charge state of an ion. We must bear in mind that the interaction with a crystal field can cause very considerable changes in the magnetic properties of T-metal ions. The Hamiltonian of a T-metal ion (4) in the presence of an external magnetic field should be supplemented by an additional term:

$$\hat{H}_H = \mu_B \vec{H}(\vec{L} + 2\vec{S}). \tag{72}$$

The dependence of the energy levels of a T-metal ion on the applied magnetic field is

$$E_i = E_i^{(0)} + E_i^{(1)}H + E_i^{(2)}H^2, \tag{73}$$

where E_i^0 is the energy of a level when the field is $H = 0$.

It is clear from Eq. (73) that an external magnetic field alters, because of the spin–orbit interaction, the system of multiplets of the crystal field. In other words, the terms split by the crystal field (see the Tanabe–Sugano diagrams in Figs. 10 and 11) experience further splitting in an external magnetic field. The fine structure of sublevels $\Gamma_1-\Gamma_n$ is formed. The magnetic susceptibility then has to be calculated from

$$\chi = N\sum_i[(E_i^{(1)})^2/kT - 2E_i^{(2)}]\exp(-E_i^{(0)}kT)/\Sigma[g_i\exp(-E_i^0/kT)], \tag{74}$$

where g_i is the multiplicity of the degeneracy of the i-th level. This formula applies to the ground state of an impurity ion when the separations between its sublevels are small compared with kT.

In those cases when it is necessary to allow for all the Γ_i sublevels in accordance with Ref. 283, the magnetic susceptibility can be expressed in terms of the crystal field parameter D_q and the spin–orbit coupling constant λ:

$$\chi = \chi_0\sum_i n_i\exp(-l_i x)/\sum_i g_i\exp(-l_i x), \tag{75}$$

where

TABLE XXX. Values of Parameters n_i and l_i Occurring in Eq. (75) (Ref. 283)

Sublevel	n_i	l_i
Γ_1	1	0
Γ_4	$\frac{1}{8}[(x+3)-3\sigma(4x+7)]$	1
Γ_3	-14σ	2
Γ_5	$\frac{1}{8}[(x-3)+\sigma(4x+21)]$	$3(1+2\sigma)$
Γ_2	$-1+4\sigma$	$4(1+3\sigma)$

Note. The notation Γ_i is explained in Fig. 92.

$$\chi_0 = 16N\mu_B^2(1+4\sigma)/K(1-5\sigma); \tag{76}$$

$$x = K(1-5\sigma)/kT; \quad K = 6\lambda^2/10D_q; \quad \sigma = \lambda/10D_q, \tag{77}$$

where the numerical values of the constants n_i and l_i of Ref. 283 are given in Table XXX.

We shall conclude this section by considering briefly the methods for the experimental determination of the magnetic susceptibility.

In the case of IV elemental semiconductors and III-V semiconductor compounds exhibiting weak magnetism, the most suitable experimental technique is the Faraday method. The basic layout of the apparatus used to measure χ by this method[284] is shown in Fig. 89.

A vacuum chamber 1 contains an electronic microbalance 2 of 1 μg sensitivity. A quartz filament 3, 150 μ in diameter, is suspended by the balance arm and this filament carries a sample 8. The pole pieces of a magnet 9 are of special shape and they create a magnetic field gradient. Glass Dewars 5 and 6 are used to carry out measurements in the temperature range 4.2-300 K. The temperature of a sample is varied by continuous pumping out a vacuum jacket 4 and by the use of a heater in the form of a bifilar winding on a copper can 7. The temperature can be maintained to within 0.05 K. Heat exchange between the copper can 7 and the sample 8 is ensured by admitting pure helium up to a pressure of 266 Pa into the chamber 1 first pumped down to 1330 Pa.

The sensitivity of the Faraday balance makes it possible to determine the magnetic susceptibility of semiconductor samples of mass amounting to a few tens of milligrammes. The absolute and relative errors do not exceed 2 and 0.5%, respectively. Samples usually have the dimensions 1.5 × 1.5 × 0.8 mm.

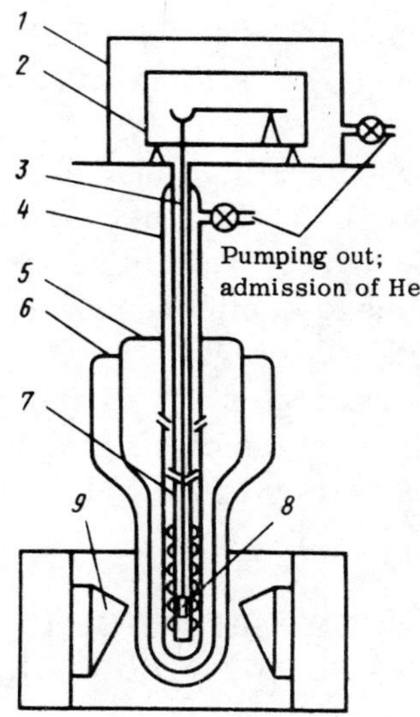

FIG. 89. Schematic diagram of the apparatus for measuring the magnetic susceptibility of semiconductors.[283]

The application of the magnetic field creates a force f_z related to dH/dz, χ, and the mass of a sample m:

$$f_z = m\chi H dH/dz, \tag{78}$$

which is used to determine χ. A direct determination of HdH/dz can be avoided by using a standard sample with a known magnetic susceptibility χ_{st}. We then have

$$\chi = (m_{st}/m)(f_z/f_{z.st})\chi_{st}. \tag{79}$$

The ratio $f_z/f_{z.st}$ is determined using the electronic balance shown in Fig. 89. The total susceptibility χ is then found from

$$\chi = \chi_{latt} + \chi_e + \chi_p. \tag{80}$$

It is shown above that at low temperatures ($T < 77$ K) the first two terms in Eq. (75) should be independent of temperature. Therefore, subtracting the constant component from the dependence $\chi(T)$, we can find χ_p due to the presence of impurities and defects. As pointed out above, the value of χ_{latt} can also be found by determining the magnetic susceptibility of an impurity-free (i.e. undoped) sample.

4.2. MAGNETIC SUSCEPTIBILITY OF GERMANIUM AND SILICON DOPED WITH T-METAL IMPURITIES

Very few experimental investigations have been made of the magnetic properties of group IV elemental semiconductors doped with transition metal impurities. This is partly due to the low solubility of T-metal atoms in these semiconductors, the maximum value of which does not exceed 10^{16} cm^{-3}.

The investigated samples[285,286] contained second-phase inclusions and the measured values of the magnetic susceptibility represented the ferromagnetism of these inclusions.

4.3. MAGNETIC SUSCEPTIBILITY OF III-V SEMICONDUCTORS DOPED WITH T-METAL IMPURITIES

Among all the III-V semiconductor compounds the fullest study of the magnetic susceptibility has been made in the case of GaAs.

Iron

An investigation of this impurity in gallium arsenide was reported in Ref. 287. The concentration of iron atoms in single crystals varied from $4 \cdot 10^{17}$ to $1 \cdot 10^{18}$ cm^{-3}.

The temperature dependence of the magnetic susceptibility (Fig. 90) revealed a strong rise of the paramagnetic contribution at low temperatures.

Calculation of χ_{latt} from the measured values of χ for samples 1 and 2 in accordance with Eq. (80) enabled the authors to determine χ_p, because the magnetic susceptibility of carriers could be ignored, since samples 1 and 2 were semi-insulating with a very low density of free holes. The value of χ_p was a sum of the paramagnetic contributions $\chi_{p1} + \chi_{p2}$ of the iron ions in the states $Fe^{3+}(3d^5)$ and $Fe^{2+}(3d^6)$, respectively.

In the case of n-type samples 3 and 4, in which the concentration of a second impurity (tellurium) exceeded the concentration of iron atoms, it was possible to use Eq. (80) to determine $\chi_{p2} + \chi_e$ and then find the susceptibility χ_{p2} since χ_e could be calculated from Eq. (63). The paramagnetic contributions of both states of the iron ions found in this way are plotted in Fig. 91.

We can see that the susceptibility of iron in the $3d^5$ state obeys the Curie law (71). This is in good agreement with the fact that the ground state ^6S of Fe^{3+} is characterized by the quantum

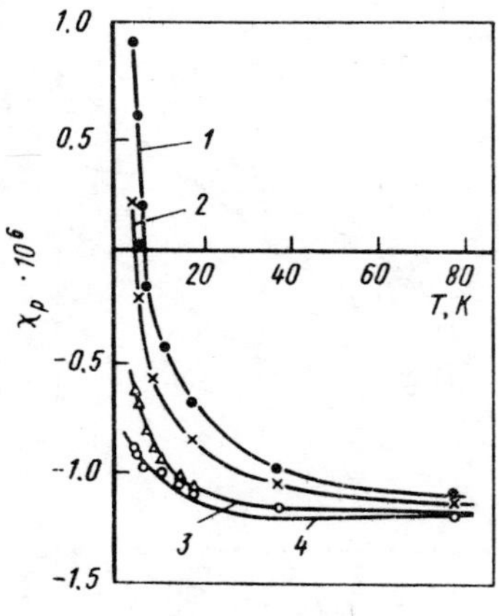

FIG. 90. Temperature dependences of the volume magnetic susceptibility of iron-doped gallium arsenide.[287] Carrier density (cm^{-3}): 1) $p = 8 \cdot 10^{12}$; 2) $p = 1 \cdot 10^{12}$; 3) $n = 2.8 \cdot 10^{17}$; 4) $n = 4.6 \cdot 10^{17}$.

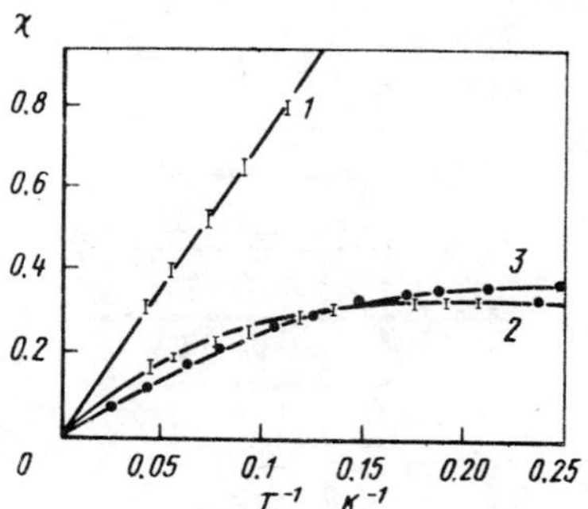

FIG. 91. Paramagnetic molar susceptibility of the iron impurity in charge states in gallium arsenide: 1) $Fe^{3+}(3d^5)$; 2) $Fe^{2+}(3d^6)$; 3) $Co^{3+}(3d^6)$. The scatter is shown for samples with different carrier densities. The continuous curves are calculated.

numbers $S = 5/2$ and $L = 0$ (Table IV) and it is not subject to the influence of the crystal field. Therefore, the magnetic properties of an ion with the electron configuration $3d^5$ should be governed by its spin magnetic moment and the magnetic susceptibility should obey the Curie law.

The temperature dependence of the magnetic susceptibility of the $Fe^{2+}(3d^6)$ ions is not described by the Curie law (Fig. 91). The ground state ^5D of the Fe^{2+} ion (Table IV) is governed, in accordance with the Hund rules, by the quantum numbers $L = 2$, $S = 2$, and $J = 4$. In the case of an ion with a nonzero orbital momentum the approximation of a free ion is no longer appropriate and in this case an analysis of the experimental results should allow for the perturbing effect of the crystal field. The splitting scheme of the energy levels of the ground state ^5D of the $Fe^{2+}(3d^6)$ ion in a crystal field of the T_d symmetry is shown in Fig. 11e. According to the calculations of Low and Weger,[288] the lower crystal-field doublet ^5E splits into a further five sublevels, the lowest of which is the singlet Γ_1 (Fig. 92). Slack et al.[283] calculated the multiplet structure of the ground term of an ion with the electron configuration $3d^6$ using higher orders of perturbation theory and they found that the energy gaps between the split levels of the ^5E doublet are not the same: in particular, the separation between the ground (lowest) singlet Γ_1 and the first excited triplet Γ_4 decreases. This

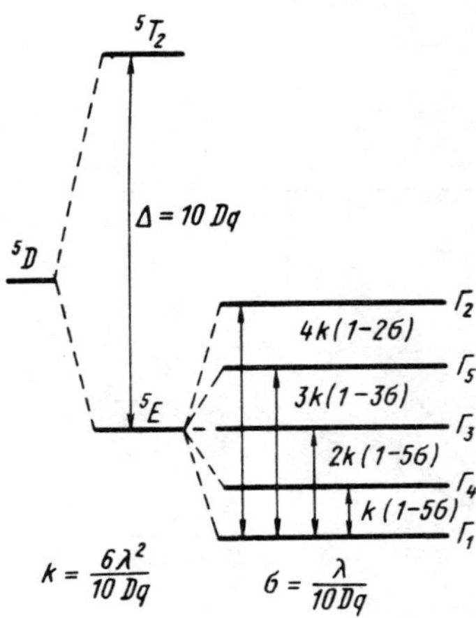

FIG. 92. Energy level scheme of T-metal ions with the $3d^6$ configuration in a tetrahedral crystal field.[283] The values of the gaps between the levels are given for Fe^{2+} on the basis of Ref. 283 and for Co^{3+} on the basis of Ref. 257. $\Gamma_5-\Gamma_2$ (cm^{-1}): 45.7 for Co^{3+} and 61.0 for Fe^{2+}. $\Gamma_3-\Gamma_5$ (cm^{-1}): 33.8 for Co^{3+} and 44.2 for Fe^{2+}. $\Gamma_4-\Gamma_3$ (cm^{-1}): 21.8 for Co^{3+} and 27.6 for Fe^{2+}. $\Gamma_1-\Gamma_4$ (cm^{-1}): 10.9 for Co^{3+} and 13.8 for Fe^{2+}.

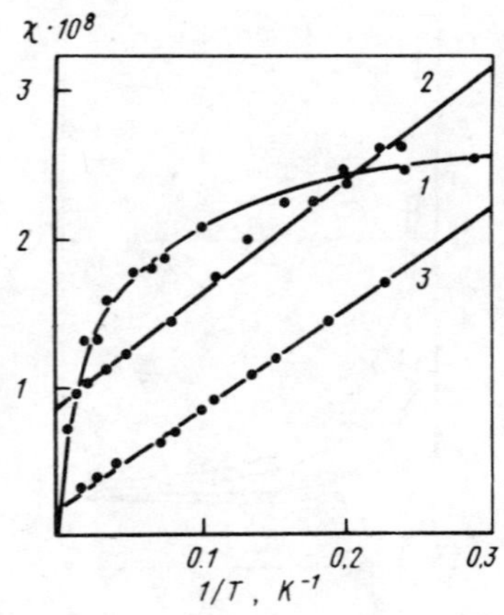

FIG. 93. Temperature dependences of the volume magnetic susceptibility of $Ni^{3+}(3d^7)$ and $Co^{2+}(3d^7)$ impurity ions in gallium arsenide (Ref. 257): 1) Ni^{3+}; 2) Co^{2+}; 3) Ni^{2+}. Curves 1 and 2 were calculated on the basis of Eq. (98).

in turn increases the Van Vleck paramagnetism of the ground state.

The experimental results (Fig. 91) are in good agreement with the theory of Slack et al.,[283] i.e. the points satisfy well the results of calculations carried out using Eqs. (75)-(77) with suitably selected parameters λ and D_q. The best agreement between the experimental results and calculations is obtained[289] for $10D_q$ = 3000 cm^{-1} and $\lambda = -90$ cm^{-1}. The splitting of the sublevels is also given in Fig. 92. The separations between the sublevels agree well with those found from the ESR and optical absorption spectra (see Sec. 1.4).

Nickel

Nickel impurities replace gallium in weakly compensated gallium arsenide samples (Sec. 1.4) and they are in the state $Ni^{3+}(3d^7)$. The temperature dependence of the magnetic susceptibility of such samples, governed by the nonionized nickel atoms (Fig. 93), is typical of the orientational Curie paramagnetism.

In fact, it follows from the appropriate Tanabe–Sugano diagram (Fig. 11f) that the lowest level under the action of the crystal field on an ion with a $3d^7$ configuration is an orbital singlet $^4A_2$. The separations between the crystal-field multiplets are fairly large (~10^4 cm^{-1}) so that at 100 K only this singlet level (characterized by a fourfold spin degeneracy) should be filled. In this case the magnetic susceptibility χ can be calculated from Eq. (74) in which the first term represents the orientational Curie paramagnetism described by the following expression in the case of an orbitally nondegenerate ground state:

$$\chi = g^2 S(S + 1)N\mu_B^2/3kT, \tag{81}$$

where $g = \sqrt{S(S + 1)}$ represents the effective magnetic moment of an ion.

The second term in Eq. (74) represents the Van Vleck polarization paramagnetism. An estimate of this paramagnetism for ions in the $3d^7$ state subject to a crystal field of the T_d symmetry was obtained by Suchkova:

$$\chi_{VV} = 8N\mu_B^2/\Delta. \tag{82}$$

A numerical estimate of the molar susceptibility χ_{VV} gives ~10^{-4}, which is less than the experimental sensitivity.

Substitution in Eq. (81) of the values of the g-factor, S (Table X), and N for nickel ions with the $3d^7$ configuration does indeed ensure a good agreement with the experimental values of the magnetic susceptibility (Fig. 93).

In the case of strongly compensated gallium arsenide samples doped with nickel and tellurium (GaAs:Ni:Te) the nickel ions are in the $Ni^{2+}(3d^8)$ state. Curve 3 in Fig. 93 is the experimental temperature dependence of the magnetic susceptibility obtained for such a sample. In the case of the Ni^{2+} ions the Curie law is not obeyed, in contrast to the Ni^{3+} ions. This can be explained by the fact that an external magnetic field creates, because of the spin–orbit interaction, a fine structure resulting in additional splitting of the sublevels, as shown in Fig. 94. We can see that the ground state $^3T_1(F)$ of Ni^{2+} splits into three sublevels, the lowest of which is a singlet A_1. Its orientational moment is zero, so that its contribution to the magnetic susceptibility appears only because of the Van Vleck paramagnetism and $\chi(T)$ should be governed primarily by the first excited state T_1. Therefore, at low temperatures, when only the ground level is populated, the magnetic susceptibility

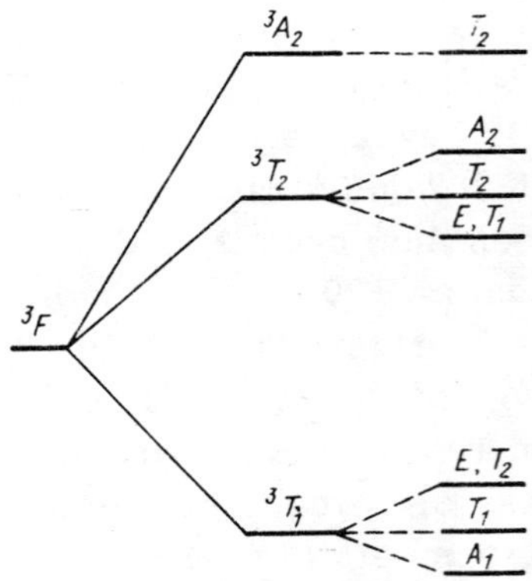

FIG. 94. Spin–orbit splitting of multiplets of a $3d^8$ ion in a crystal field of the T_d symmetry.[290]

should be independent of the temperature, as manifested by the tendency of curve 3 in Fig. 93 to saturation.

Cobalt

This impurity is found in the $3d^7(Co^{2+})$ state in gallium arsenide only if samples are strongly compensated. We can demonstrate this by considering the ESR spectra (Sec. 1.4). Therefore, the temperature dependence of the magnetic susceptibility of GaAs:Co: Te is of the same form as that of GaAs:Ni (Fig. 93). Once again a good agreement between the experimental results and theory is obtained when Eq. (81) is used and the values of S and g are substituted from Table X.

It should be pointed out that these samples manifest not only the magnetism of Co^{2+} ions but also an unidentified constant paramagnetic contribution, which is demonstrated clearly in Fig. 93. This contribution cannot be explained by the presence of small amounts of a foreign impurity (iron) nor by the presence of cobalt in any other state. The origin of this small additional paramagnetic contribution exhibited by cobalt-doped gallium arsenide is thus still unclear. The trivalent Co^{3+} cobalt ions should behave fully analogously to the Fe^{2+} ions, because both of them have the $3d^6$ configuration in gallium arsenide. Samples of this kind can be prepared by simultaneous introduction of cobalt and a shallow acceptor (zinc) in an amount sufficient for the compensation of the residual donor background in GaAs, which ensures decompensation of the cobalt impurity levels. In samples of this kind the cobalt ions are in the nonionized $Co^{3+}(3d^6)$ state.

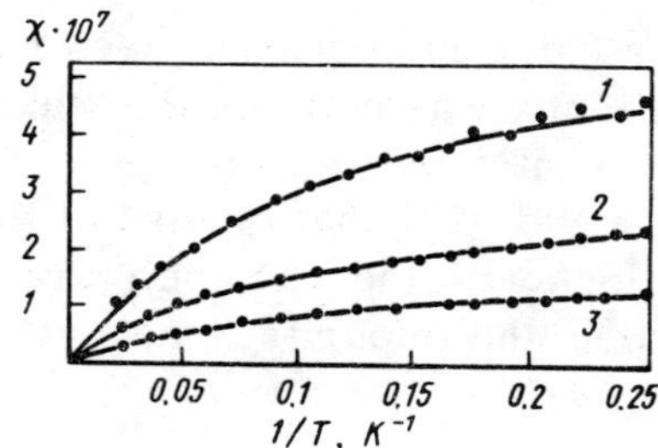

FIG. 95. Temperature dependences of the volume magnetic susceptibility of the $Co^{3+}(3d^6)$ ions in gallium arsenide.[257] The continuous curves represent calculations. Concentration of Co^{3+} ions (cm^{-3}): 1) $7\cdot10^{17}$; 2) $3.5\cdot10^{17}$; 3) $1.8\cdot10^{17}$.

This was confirmed experimentally (Fig. 95). In this case the temperature dependence $\chi(T)$ also agrees well with theory when calculations are made using Eqs. (75)-(77) and the values $10D_q$ = 10 000 cm^{-1} and $\lambda = -145$ cm^{-1}. The splitting of the sublevels[257] is also given in Fig. 92. Compared with $Fe^{2+}(3d^6)$, the separations between the sublevels of the multiplet ^5E of the Co^{3+} ion are smaller. This results in a plateau of the dependence $\chi(T)$ at lower temperatures, in good agreement with the experimental results (see Fig. 91).

Chromium

This metal is an impurity with a magnetic moment in III-V crystals and this is undoubtedly of considerable interest. The crystal field is of the T_d symmetry and it splits the terms (Fig. 35), so that in the case of $Cr^{2+}(3d^4)$ the dynamic Jahn–Teller effect is active. It is stated in Ref. 48 that this effect is not strong and it does not alter qualitatively the structure of the electron spectrum, whereas according to Ref. 56 the Jahn–Teller distortions are considerable and they lower the symmetry of the $Cr^{2+}(3d^4)$ centre.

This conflict cannot be resolved by investigating optical intracentre transitions, since the upper split-off state lies within the allowed spectrum (Sec. 3.3) and the fine structure of the spectra is not resolved.

Magnetic measurements should definitely be sensitive to the Jahn–Teller effect. However, these measurements are difficult to carry out because in wide-gap III-V semiconductors (for example, GaAs) the chromium impurity forms multiply charged centres and the Cr^{2+} state can be achieved only when the concentration of chromium is exactly equal to that of shallow donors. This is almost impossible to realise experimentally. A deviation in one or the other direction away from this equality gives rise to the Cr^{3+} or Cr^{+} ions. This difficulty was noted in the magnetic investigations of GaAs:Cr and was pointed out in Ref. 291.

It was shown in Ref. 291 that studies of the magnetic properties of the T-metal ions with the $3d^4$ configuration are best carried out using manganese as the impurity.

Manganese

Investigations of this impurity in InAs, GaAs, and InP were reported in Ref. 291. The freezeout of holes occurs in all three matrices at low temperatures. Consequently, the manganese ions are in the state $Mn^{3+}(3d^4)$.

We shall consider the behaviour of manganese in InAs:Mn, because it was analyzed in Ref. 291. Temperature dependences of the magnetic susceptibility reported in that paper are reproduced in Fig. 96.

At high temperatures the manganese ions are in the ionized

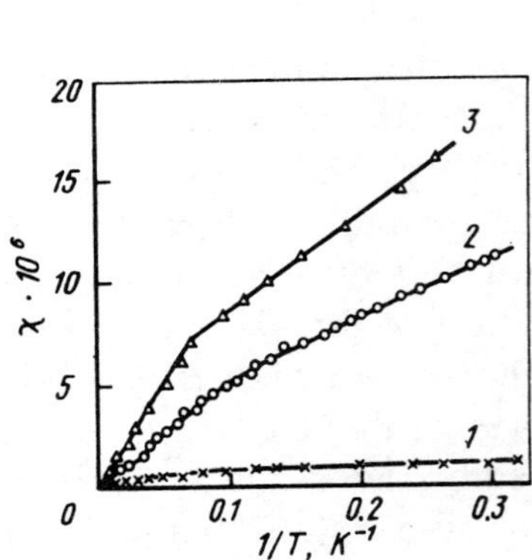

FIG. 96. Temperature dependences of the volume magnetic susceptibility of manganese impurity ions in InAs (Ref. 291). Manganese concentration (10^{18} cm^{-3}): 1) 0.45; 2) 4.9; 3) 84.

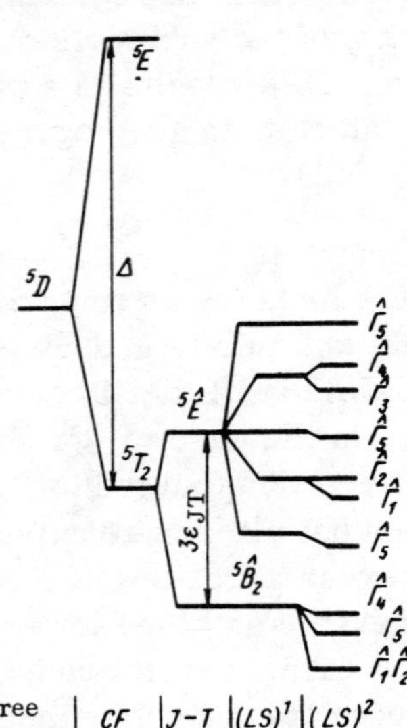

FIG. 97. Splitting of the energy levels of a $3d^4$ ion under the influence of a crystal field of the T_d symmetry (CF in the figure), of Jahn–Teller deformations (ε_{JT}) along the ⟨100⟩ directions, and of spin–orbit interaction of the first $(LS)^1$ and second $(LS)^2$ orders.

state $Mn^{2+}(3d^5)$. It should be noted that the ions with the $3d^5$ configuration are not affected by the crystal field and their magnetic susceptibility is described by the Curie law (81).

It is shown in Ref. 291 that substitution of $S = 5/2$ and $g = 2$ (Tables IV and X) ensures an agreement between the value of N calculated from Eq. (81) and the density of holes deduced from the Hall coefficient. This is a further confirmation that the presence of manganese ions at high temperatures results in a contribution of one hole to the valence band.

The low-temperature part of the dependence $\chi(T)$ is governed entirely by the $Mn^{3+}(3d^4)$ ions. According to the term splitting scheme of the $3d^4$ ion (Fig. 35b), the lowest state is a complete singlet. Estimates obtained in Ref. 291 indicated a gap between this state and the first excited level is such that the majority of the ions are in the ground state at liquid helium temperature. Therefore, we can expect that, as in the case of the $Fe^{2+}(3d^6)$, $Ni^{2+}(3d^8)$, and $Co^{3+}(3d^6)$ ions in GaAs discussed above, the magnetic susceptibility of the $Mn^{3+}(3d^4)$ ions in InAs should exhibit a plateau at low temperatures. However, the curves plotted in Fig. 96 show no such plateau. This is explained in Ref. 291 by the static Jahn–Teller effect. The energy scheme obtained in this case is shown in Fig. 97. If we assume that the magnitudes of the Jahn–Teller deformations, i.e. $3\varepsilon_{JT}$, in III-V compounds are of the same order of magnitude as in II-VI compounds (in the latter case it is reported in Ref. 292 that $3\varepsilon_{JT} = 1500\ cm^{-1}$), then only the structure of a multiplet associated with the orbital singlet ${}^5\hat{B}_2$ is important (Fig. 97).

It is also shown in Ref. 291 that at 4.2 K the majority of the $Mn^{3+}(3d^4)$ ions occupies the lower split-off state $\hat{\Gamma}_1\hat{\Gamma}_2$ shown in the above scheme.

Bearing in mind that the projection of the spin momentum along the $\vec{H}$ direction assumes in this state only two values $S_z = \pm 2$, the authors of Ref. 291 found the magnetic susceptibility of those ions for which the direction of the Jahn–Teller distortions coincides with the vector $\vec{H}$:

$$\chi = ng^2\mu_B^2S_z^2/kT. \tag{83}$$

Unfortunately, the investigations reported in Ref. 291 were carried out on unoriented samples, in which case the dependence $\chi(T)$ should be described by Eq. (83) but with some average value of $g^2S_z^2$. A comparison of Eqs. (81) and (83) shows that the ratio of the slopes of the two rectilinear parts of the curves in Fig. 96 is

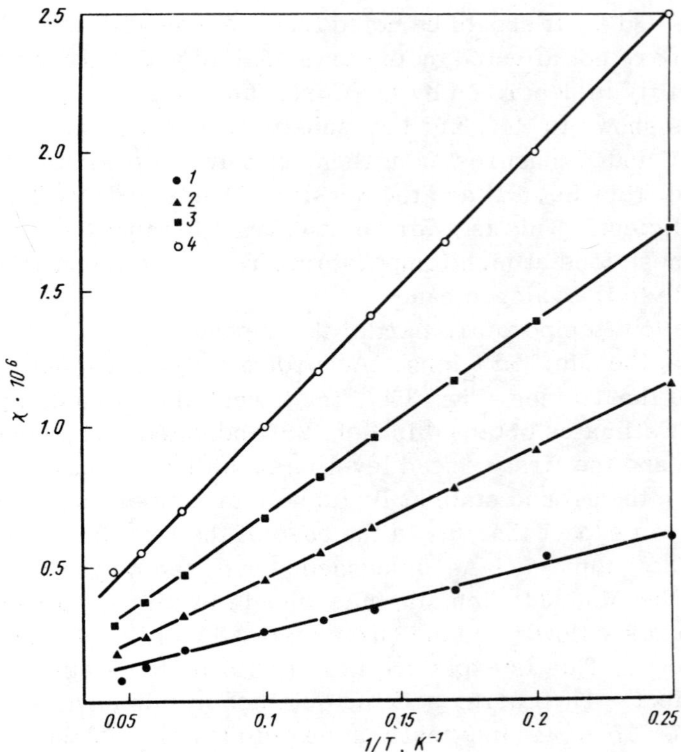

FIG. 98. Temperature dependences of the volume magnetic susceptibility of manganese ions in InSb (Ref. 293). Hole density at 300 K (cm^{-3}): 1) $3\cdot10^{17}$; 2) $6\cdot10^{17}$; 3) $1\cdot10^{18}$; 4) $1.3\cdot10^{18}$.

$$f = g_0^2 S_0 (S_0 + 1)/3g^2 S_z^2, \tag{84}$$

where g_0 and S_0 are obtained at high temperatures.

A calculation carried out using Eq. (84) gives $f = 2.2$, whereas the experimental values are $f = 2.1$-2.4 (Ref. 291). Such an agreement confirms the validity of an allowance for the Jahn–Teller effect in the specific case of the T-metal ions in the $3d^4$ configuration in III–V compounds.

It is shown in Ref. 293 that manganese in InSb is in the $3d^5$ state, since the energy gap between the valence band and the acceptor level of manganese is so small that it vanishes already for 10^{17} cm^{-3}.

The $3d^5$ ion is characterized by zero orbital momentum and, therefore, in the first approximation of perturbation theory, the $Mn^{2+}(3d^5)$ ion is not subject to the influence of the crystal field. The magnetic moment of such an ion is governed by its spin $S = 5/2$ (Table IV) and is (for $g = 2$):

$$\mu = g\mu_B\sqrt{S(S+1)} = 5.92\,\mu_B, \tag{85}$$

and the dependence $\chi(T)$ should be described by the Curie law (81), which is indeed observed experimentally (Fig. 98).

4.4. NEGATIVE MAGNETORESISTANCE AND OTHER MAGNETIC EFFECTS IN SEMICONDUCTORS DOPED WITH T-METAL IMPURITIES

The negative magnetoresistance occurs at low temperatures and is manifested by a reduction in the resistivity ρ on application of an external magnetic field. Typical dependences of $\Delta\rho/\rho$ on the magnetic field intensity are shown in Fig. 99.

It should be pointed out that the usual theory of galvanomagnetic effects is based on a model in which an external magnetic field bends at the trajectory of an electron. The resultant Lorentz force drives an electron along a circular orbit about the vector $\vec{H}$. Consequently, the resistivity should increase because the electron trajectory becomes longer and, consequently, the number of positions experienced by the electron becomes large, in other words the mean free path of an electron decreases and so does its mobility. Therefore, the electrical resistivity should rise on increase in H in accordance with the law $\rho \propto H^2$.

It is clear from Fig. 99 that at high values of H the dependence of ρ on H^2 is indeed linear, but this is not true in low fields

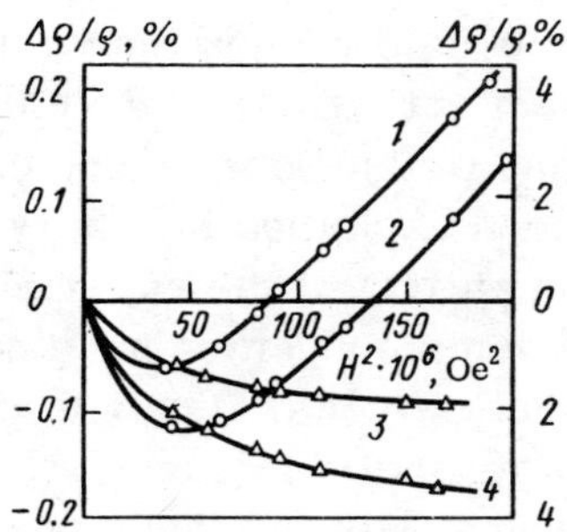

FIG. 99. Magnetoresistance of germanium samples: 1), 2) Ge:As, $n = 5.2 \cdot 10^{17}$ cm^{-3}; 3), 4) Ge:Sb, $n = 6 \cdot 10^{17}$ cm^{-3}.

where ρ decreases. The experimentally observed curve can be divided into two components: the normal positive Lorentzian and the anomalous negative, the latter rising in the absolute sense on increase in the field and then reaching saturation.

The negative magnetoresistance of semiconductors is explained[294] by the scattering of conduction electrons on localized magnetic moments. This idea was borrowed from Ref. 295, where a theory of the negative magnetoresistance is developed for nonmagnetic metals doped with transition-element atoms exhibiting a magnetic moment. According to this theory the negative magnetoresistance should be proportional to the square of the magnetization I, i.e. $\Delta\rho/\rho \propto I^2$, as indeed found experimentally for metals representing dilute solutions of magnetic impurities, for example Cu–Mn.

If the magnetization of a sample is described by the Brillouin function $B_J(\mu H/kT)$, where μ is the magnetic moment of the scattering centres, or if it is described by its classical analogue in the form of the Langevin function $L(\mu H/kT)$, the negative magnetoresistance should be described by

$$\Delta\rho/\rho \propto B_J(\mu H/kT) \tag{86}$$

or

$$\Delta\rho/\rho \propto L^2(\mu H/kT). \tag{87}$$

An analysis of the experimental data on the negative magnetoresistance of semiconductor crystals[296–298] shows that Eqs. (86) and (87) are satisfied only if we use a magnetic moment μ which is dependent on the magnetic field. Then the values of the magnetic moments found by different authors are fairly large and range from 4-5 μ_B to 20-30 μ_B (μ_B is the Bohr magneton). However, neither of these results has been explained in a fully satisfactory manner.

Moreover, the majority of the experiments on the negative magnetoresistance of semiconductors have been carried out on crystals doped with shallow donor impurities, singly ionized atoms of which do not have a magnetic moment. Therefore, some authors have suggested various models for the appearance of the magnetic moment in such crystals.

The main models postulate localization of an electron at a "solitary" donor atom, the probability of the appearance of which is related to the random distribution of impurities,[296] partial localization of an electron in an impurity energy band,[294,298] localization of an electron in a potential well formed as a result of corrugation of the bottom of the conduction band in compensated crystals,[297,299] and finally the hypothesis[300] that carriers of localized spin in heavily doped samples are some impurity atoms which are also responsible for the disagreement between the chemical concentration of impurities and the Hall density of conduction electrons found in the case of heavily doped semiconductors. This mechanism is supported also by a more direct evidence obtained by measurements of the magnetic susceptibility of samples which do not contain T-metal ions.

It is natural to expect each of the proposed models of the localization of the magnetic moment to be realised with the highest probability in a certain range of the dopant concentrations. Each of the reasons for the localization should give rise to different magnetic moments. This may be due to different values of the Landé factor g and due to different values of the total quantum number J of the localized magnetic moment. An increase in J on increase in the impurity concentration is exhibited, for example, by samples of a dilute solution of manganese in lithium[301] and is explained by the formation of magnetic clusters.

Hence it follows that each specific crystal may be characterized by a set of different magnetic moments and in this case its magnetization cannot be described by the Brillouin or Langevin function.

A calculation of the magnetization of such a crystal is difficult because the quantity I, determined by each pair of the values of g and J, should be found from a sum of the magnetizations due to all possible mechanisms of the localization of the magnetic moment. For each specific localization mechanism we can expect creation of the magnetic moment within certain limits. Therefore, in calculations of the magnetization ensured by a specific localization mechanism, we should use not the exact values of g and J, but the average of all the possible values of g and J taken with suitable statistical weights. Under these conditions the negative magnetoresistance cannot be described by Eqs. (86) or (87), and attempts to compare the experimentally determined dependences $\Delta\rho/\rho = f(H, T)$ with the Brillouin or Langevin functions unavoidably lead to the introduction of a magnetic moment which depends both on temperature and on the applied magnetic field.[296–299]

In general, the dependence of the magnetic moment on T and H applies also to magnetic centres of the same kind. For example, in the case of iron impurity atoms in gallium arsenide doped simultaneously with iron and tellurium so that $N_{Te} > N_{Fe}$, the iron ions have the electron configuration $3d^6$ and at 4.2 K these ions have magnetic centres with a polarization magnetic moment (Sec. 4.3). The negative magnetoresistance of these samples is plotted in Fig. 100. We can see that doping with iron increases considerably the negative magnetoresistance compared with samples with a similar electron density, but doped with tellurium alone. Similar results were obtained also for GaAs:Cr samples.[303]

However, the conclusion that the increase in the magnetoresistance is in both cases entirely due to the presence of a magnetic impurity is of doubtful validity because in both cases we are dealing with compensated samples and compensation can also give rise to a negative magnetoresistance.

The interpretation of the observed rise of the negative magnetoresistance of gallium arsenide doped with iron and tellurium (Fig. 100) is difficult also because according to the theory of the crystal field and our experimental investigations of the magnetic susceptibility of such crystals (Sec. 4.3), the orientational magnetic moment of the iron ions should tend to zero as temperature is lowered.

Therefore, we investigated the negative magnetoresistance of InSb:Mn (Ref. 293). In this case the manganese atoms not only ensure a high density of carriers, which makes it easier to observe the negative magnetoresistance, but they also have a magnetic moment which is independent of temperature.

At low temperatures the negative magnetoresistance effect (Fig. 101) is due to the spin scattering of holes by magnetic moments of the manganese ions. This conclusion is confirmed by the observation that doping of InSb with the zinc and cadmium impurities, which do not have a magnetic moment, gives rise to a magnetoresistance which is positive throughout the investigated range of magnetic fields (Fig. 101).

According to the theoretical calculations,[294,295] the observed negative magnetoresistance should be described by Eq. (86) if $J = 5/2$. However, our analysis of the experimental results shows that Eq. (86) cannot explain the dependences of the negative magnetoresistance of InSb:Mn samples on T and H at temperatures 2-10 K and in magnetic fields up to 10 kOe.

It is interesting to note that a dependence $S^{-1/2} = f(T)$, where S is a quantity defined as follows[294]

$$S = \lim_{H \to 0} [-\rho(H) - \rho(0)]/\rho(0)H^2 \qquad (88)$$

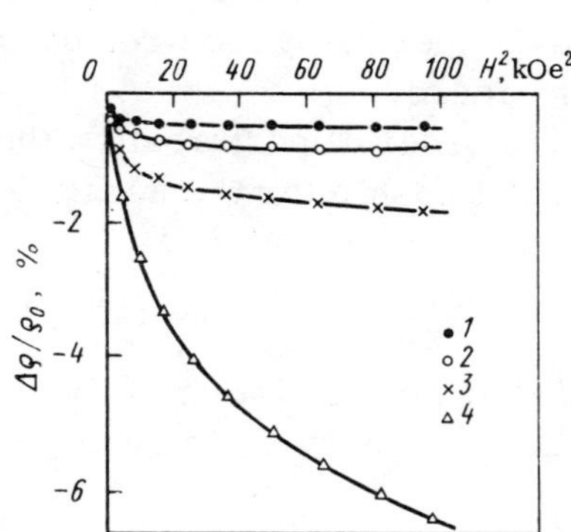

FIG. 100. Longitudinal magnetoresistance of gallium arsenide samples at 4.2 K (Ref. 302): 1) GaAs:Te, n = $4 \cdot 10^{17}$ cm^{-3}; 2) GaAs:Te, n = $2.6 \cdot 10^{17}$ cm^{-3}; 3) GaAs:Te:Fe, n = 4.10^{17} cm^{-3}; 4) GaAs:Te:Fe, n = $2 \cdot 10^{17}$ cm^{-3}. Iron concentrations $7 \cdot 10^{17}$ cm^{-3} (3) and $6.4 \cdot 10^{17}$ cm^{-3} (4).

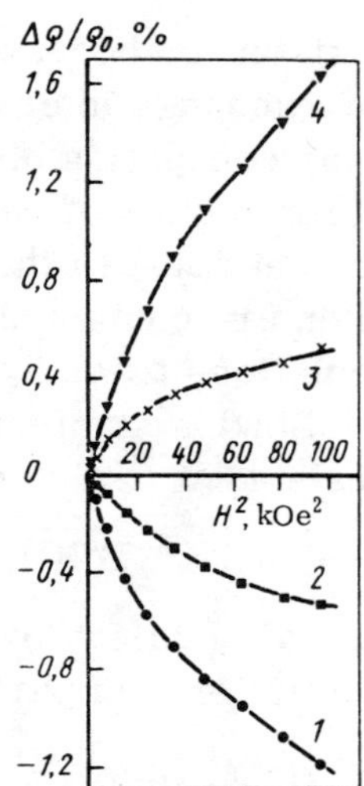

FIG. 101. Longitudinal magnetoresistance of p-type InSb samples at 4.2 K (Ref. 293). Dopant and hole density (cm^{-3}); 1) Mn, $3 \cdot 10^{17}$; 2) Mn, $1 \cdot 10^{18}$; 3) Cd, $5.6 \cdot 10^{18}$; 4) Zn, $1.3 \cdot 10^{18}$.

for the investigated InSb:Mn samples at temperatures 2.6 K, is analogous to the Curie–Weiss law with a negative value of the characteristic temperature θ. A similar analysis of the negative magnetoresistance data obtained for other semiconductors also gives negative values of θ. However, this is in conflict with the results of an investigation of the magnetic susceptibility of the same InSb:Mn samples, since the value of χ obeys the Curie law when $\theta = 0$ (Fig. 98).

It is worth noting another circumstance. According to Eq. (86), an increase in the concentration of localized magnetic moments should increase the negative magnetoresistance. However, it was found that an increase in the manganese concentration not only did not enhance the negative magnetoresistance, but in fact it reduced this effect (Fig. 101). This was evidently associated with the dependence of the strength of the exchange interaction on the relative position of the Fermi level and of the impurity centre on the energy scale, as pointed out in Refs. 293 and 302.

Khosla and Fischer[304] concluded that Eq. (86) does not represent the temperature and magnetic-field dependences of the negative magnetoresistance, because it was obtained as a result of a calculation of the exchange interaction between the localized magnetic moments and conduction electrons (S–d interaction) only in the second approximation of perturbation theory. In their opinion, the best agreement between the theory and experiment requires calculations which are continued to higher orders.

Calculations extended to the third order of perturbation theory led to the following semi-empirical relationship for the negative magnetoresistance[304]:

$$\Delta\rho/\rho = -B_1^2 \ln(1 + B_2^2 H^2), \tag{89}$$

where B_1 and B_2 allow for the physical characteristics of the exchange interaction. In particular, the coefficient B_2 is given by

$$B_2^2 = [1 + 4S^2\pi^2(2J\rho_F/q^4)]g^2\mu_B^2/(\alpha kT)^2, \tag{90}$$

where J is the exchange interaction energy; ρ_F is the density of states at the Fermi level; α is a numerical coefficient which can be quite satisfactorily assumed to be unity; S and g are the spin and Landé factor of a localized magnetic moment.

An analysis of the experimental results in Fig. 101 (Ref. 293) showed that they were in good agreement with Eq. (89). However, since this equation is semi-empirical, a better understanding of the physical processes responsible for the negative magnetoresistance effect in semiconductors will require further theoretical studies.

It is worth noting a theory of quantum corrections[304a] to the usual conductivity based on the wave nature of the electron scattering. In this theory the temperature and magnetic-field dependences of the negative magnetoresistance do not require the existence of localized magnetic moments. A review of the experimental data in the light of the new negative magnetoresistance theory would undoubtedly be of interest.

Among the other magnetic effects which appear in semiconductors when they are doped with heavy-metal impurities one should mention particularly the anomalous Hall effect.

This effect appears when a current $\vec{j}$ passes through a magnetized sample and it is manifested externally by an electric field $\vec{E}$ directed at right-angles to the vector $\vec{j}$ and to the magnetization vector $\vec{I}$. It should be stressed that, in contrast to the normal Hall effect, the anomalous effect appears only in ferromagnets and paramagnets, and its e.m.f. is proportional to the magnetization of the sample. A similar effect was reported for InSb:Mn samples.[305]

CONCLUSIONS

We shall now draw some conclusions. The reported investigations of deep centres in the specific case of transition metal impurities in germanium, silicon, and some III-V compounds taken as a whole have made it possible to extend our basic ideas on the centres of strong localizations in semiconductors, specific nature of transition metal impurities, and their influence on a great variety of crystal properties. The results obtained are also of considerable practical interest in connection with a possibility of controlled modification of the properties of semicondutor crystals in a very wide range of values of the principal parameters so that it should be possible to create an "optimal" material for each specific application. In particular, use of the data on the state and behaviour of transition metal impurities in gallium arsenide has made it possible to develop a reliable technology for commercial preparation of semi-insulating GaAs single crystals and later of other III-V compounds, the applications of which in solid-state electronics are steadily rising.

However, not everything is clear about the properties of the materials of interest to us which are doped with impurities that have partly filled 4d-, 5d-, or *f*-shells with certain specific features. In particular, it is of interest to consider the problem of the s–*f* interaction which should clearly result in the formation of deep local states, as already done on the basis of the Anderson model for the s–d hybridization.

In considering the need to develop a theory of deep centres, including T-metal impurities, it is convenient to list those theoretical topics to which experimentalists are awaiting answer. In our opinion, a sufficiently comprehensive theory should describe a wide range of the properties of deep centres, including:

the nature of deep levels created by impurities or defects;

the perturbation potential introduced by a defect allowing for its individual properties;

the structure and symmetry of local states in the band scheme of crystals found allowing for the main forms of perturbation of the energy spectrum, including the electron–phonon interaction and lattice relaxation;

the spectral dependence of the photon absorption cross-section
the fine structure of the ESR and ENDOR spectra;
the carrier-scattering and capture cross-sections of deep levels.

The reported investigations have been concentrated mainly on bulk crystals containing transition metal impurities. However, the gradual change to epitaxial films has resulted in a considerable improvement of the quality of many devices made of high-resistivity semiconductors. The validity of extrapolation of the available data on the properties of T-metal impurities in bulk crystals to epitaxial films is in general doubtful because the state of impurities in the crystal lattice may depend strongly on the method and conditions of growth of a material. It would seem that the published data on the nature of the epitaxial GaAs films containing iron[306] or manganese[307] impurities would allow the conclusion that their properties are identical with those of the bulk materials. However, the results of an investigation of the parameters of GaAs:Cr films prepared by the vapour[308] or liquid[309] epitaxy raise doubts about the identity of the properties of all the transition-metal impurities in the bulk material and in epitaxial GaAs films.

Recent papers contain reports of determination of the local symmetry of the chromium impurity ions in the GaAs and GaP lattices on the basis of a detailed analysis of the fine structure of low-temperature photoluminescence and cathodoluminescence spectra, ESR and PAR spectra, and other data allowing for the Jahn–Teller effect and for lattice relaxation near an impurity centre (see, for example, Refs. 310 and 311). The results of these investigations are highly contradictory and this may be particularly due to the various previous histories of the investigated samples. Therefore, a comprehensive investigation carried out on the same crystals would have allowed us to draw more definite conclusions. There are grounds for assuming that in both GaAs and GaP some of the chromium atoms interact with other structure defects, for example, group VI donors or with oxygen, which is the main residual impurity in these materials. Consequently, the configuration of a complex formed from an impurity atom and its immediate environment experience fairly strong distortions and this reduces the symmetry of the centres and splits the energy levels accordingly. However, the validity of this approach must be checked by further investigations and it is necessary to determine whether this interaction is a characteristic property of all transition-metal impurities in various matrices or it is solely due to chromium atoms in GaAs and GaP crystals. Further pursuit of these investigations is of major scientific and practical importance.

The problems not yet fully resolved are as follows:

1) the local symmetry and the positions occupied by chromium atoms and other transition metals in narrow-gap III-V compounds;

2) the localization radius of the wave function of impurity atoms belonging to the iron group (this problem should be tackled by a quantitative interpretation of the ENDOR spectra);

3) a detailed allowance for the electron–phonon interaction with a small-radius impurity centre, almost ignored completely, but manifested, for example, in the optical absorption spectra in the region of the charge transfer band and of the intracentre transitions, and also in ESR and PAR.

This is not a complete list of tasks which are facing investigators. However, we can expect further growth of experimental and particularly theoretical investigations which should make it possible to solve the "old" and still topical problem of deep centres in semiconductors.

BIBLIOGRAPHY

1. S.S. Batsanov, Electronegativity of Elements and Chemical Binding [in Russian], Izd. SO AN SSSR, Novosibirsk (1962).
2. W.P. Pearson, Crystal Chemistry and Physics of Metals and Alloys, Wiley, New York (1972).
3. N.N. Sirota, Physicochemical Nature of Variable-Composition Phases [in Russian], Nauka i Tekhnika, Minsk (1970).
4. N.S. Hush and M.H.L. Pryce, J. Chem. Phys. 28, 244 (1958).
5. I.B. Bersuker, Electron Structure and Properties of Coordination Compounds [in Russian], Khimiya, Leningrad (1976).
6. A.S. Marfunin, Introduction to Physics of Minerals [in Russian], Nedra, Moscow (1974).
7. S. Sugano, Y. Tanabe, and H. Kamimura, Multiplets of Transition-Metal Ions in Crystals, Academic Press, New York (1970).
8. J.M. Baranowski, J.W. Allen, and G.L. Pearson, Phys. Rev. 167, 758 (1968).
9. A.M. Hennel, Phys. Status Solidi B 72, K9 (1975).
10. R. McWeeny and B.T. Sutcliffe, Methods of Molecular Quantum Mechanics, Academic Press, New York (1969).
11. Electron Spin Resonance in Semiconductors [Russian translation ed. by N.A. Penin], IL, Moscow (1962).
12. V.S. Vavilov, O.G. Koshelev, and Yu.P. Koval', Fiz. Tekh. Poluprovodn. 1, 15 (1967) [Sov. Phys. Semicond. 1, 10 (1967)].
13. H.H. Woodbury and G.W. Ludwig, Phys. Rev. 117, 102 (1960).
14. W. Low, Paramagnetic Resonance in Solids, Suppl. No. 2 to Solid State Phys., Academic Press, New York (1960).
15. G.W. Ludwig and H.H. Woodbury, Solid State Phys. 13, 223 (1962).
16. B. Bleaney and R.S. Rubins, Proc. Phys. Soc. London 77, 103 (1961).
17. J.S. Griffith, Proc. R. Soc. London Ser. A 235, 23 (1956).
18. J.E. Geusic, M. Peter, and E.O. Schulz-DuBois, Bell Syst. Tech. J. 38, 291 (1959).
19. R.L. Mössbauer, Usp. Fiz. Nauk 72, 658 (1960) [Sov. Phys. Usp. 3, 866 (1960)].
20. V.I. Goldansky, Mössbauer Effect and Its Applications to Chemistry, Van Nostrand Reinhold, New York (1966).
21. Yu. Kagan and V.A. Maslov, Zh. Eksp. Teor. Fiz. 41, 1296 (1961) [Sov. Phys. JETP 14, 922 (1962)].
22. Paramagnetic Resonance (Proc. First Intern. Conf., Jerusalem, 1962, ed. by W. Low), 2 vols., Academic Press, New York (1963).
23. W. Fahrner and A. Goetzberger, Appl. Phys. Lett. 21, 329 (1972).
24. H.H. Woodbury and G.W. Ludwig, Phys. Rev. Lett. 5, 96 (1960).
25. G.D. Watkins, Bull. Am. Phys. Soc. 2, 345 (1957).
26. W. Low and M. Weger, Phys. Rev. 118, 1119, 1130 (1960).
27. G. Feher, Phys. Rev. 114, 1219 (1959).
28. T. Igo, Rev. Elec. Commun. Lab. 10, 300 (1962).
29. N.T. Bendik, L.S. Milevskiĭ, and E.G. Smirnov, Fiz. Tekh. Poluprovodn. 5, 848 (1971) [Sov. Phys. Semicond. 5, 749 (1971)].
30. G.W. Ludwig, H.H. Woodbury, and R.O. Carlson, Phys. Rev. Lett. 1, 295 (1958).
31. G.W. Ludwig and H.H. Woodbury, Phys. Rev. 117, 1286 (1960).
32. G. Lancaster, Electron Spin Resonance in Semiconductors, Hilger & Watts Ltd., London (1966).
33. G.W. Ludwig, R.O. Carlson, and H.H. Woodbury, Bull. Am. Phys. Soc. 4, 22, 144 (1959).

34. G.W. Ludwig and H.H. Woodbury, Phys. Rev. 113, 1014 (1959).
35. H.H. Woodbury and G.W. Ludwig, Phys. Rev. Lett. 1, 16 (1958).
36. A.B. Roĭtsin and L. A. Firshteĭn, Teor. Eksp. Khim. 2, 747 (1966).
37. E. Clementi and D.L. Raimondi, J. Chem. Phys. 38, 2686 (1963).
38. G.W. Ludwig and H.H. Woodbury, Phys. Rev. Lett. 5, 98 (1960).
39. G.W. Ludwig, H.H. Woodbury, and R.O. Carlson, J. Phys. Chem. Solids 8, 490 (1959).
40. D.G. Andrianov, N.M. Pavlov, A.S. Savel'ev, V.I. Fistul', and G.P. Tsiskarishvili, Fiz. Tekh. Poluprovodn. 14, 1202 (1980) [Sov. Phys. Semicond. 14, 711 (1980)].
41. M. De Wit and T.L. Estle, Bull. Am. Phys. Soc. 7, 449 (1962).
42. R. Bleekrode, J. Dieleman, and H.J. Vegter, Philips Res. Rep. 17, 513 (1962).
43. B. Goldstein and N. Almeleh, Appl. Phys. Lett. 2, 130 (1963).
44. V.K. Bazhenov, V.A. Presnov, and S.P. Fedotov, Fiz. Tverd. Tela (Leningrad) 10, 268 (1968) [Sov. Phys. Solid State 10, 205 (1968)].
45. V.K. Bashenov, Phys. Status Solidi A 10, 9 (1972).
46. N. Almeleh and B. Goldstein, Phys. Rev. 128, 1568 (1962).
47. V.I. Fistul', L.Ya. Pervova, É.M. Omel'yanovskiĭ, E.P. Rashevskaya, N.N. Solov'-ev, and O.V. Pelevin, Fiz. Tekh. Poluprovodn. 8, 485 (1974) [Sov. Phys. Semicond. 8, 311 (1974)].
48. E.M. Ganapol'skiĭ, Fiz. Tverd. Tela (Leningrad) 15, 368 (1973) [Sov. Phys. Solid State 15, 269 (1973)].
49. E.M. Ganapol'skiĭ, É.M. Omel'yanovskiĭ, L.Ya. Pervova, and V.I. Fistul', Fiz. Tekh. Poluprovodn. 7, 1643 (1973) [Sov. Phys. Semicond. 7, 1099 (1974)].
50. J.M. Baranowski, J.W. Allen, and G.L. Pearson, Phys. Rev. 160, 627 (1967).
51. G.K. Ippolitova and É.M. Omel'yanovskiĭ, Fiz. Tekh. Poluprovodn. 9, 236 (1975) [Sov. Phys. Semicond. 9, 156 (1975)].
52. F.S. Ham, Phys. Rev. 138, A1727 (1965).
53. G.N. Belozerskiĭ, Yu.A. Nemilov, S.B. Tomilov, and A.V. Shvedchikov, Fiz. Tverd. Tela (Leningrad) 7, 3607 (1965) [Sov. Phys. Solid State 7, 2908 (1966)].
54. D.G. Andrianov, A.S. Savel'ev, and V.I. Fistul', Fiz. Tekh. Poluprovodn. 9, 136 (1975) [Sov. Phys. Semicond. 9, 89 (1975)].
55. W. Teuerle and A. Hausmann, Z. Phys. B 23, 11 (1976).
56. J.J. Krebs and G.H. Stauss, Phys. Rev. B 15, 17 (1977).
57. B. Frick and D. Siebert, Phys. Status Solidi A 41, K185 (1977).
58. E.M. Ganopol'skiĭ, Fiz. Tverd. Tela (Leningrad) 16, 2886 (1974) [Sov. Phys. Solid State 16, 1868 (1975)].
59. G.K. Ippolitova, É.M. Omel'yanovskiĭ, and L.Ya. Pervova, Fiz. Tekh. Poluprovodn. 9, 1308 (1975) [Sov. Phys. Semicond. 9, 864 (1975)].
60. G. Lucovsky, Bull. Am. Phys. Soc. 11, 206 (1966).
61. J.T. Vallin, G.A. Slack, S. Roberts, and A.E. Hughes, Phys. Rev. B 2, 4313 (1970).
62. U. Fano and J.W. Cooper, Rev. Mod. Phys. 40, 441 (1968).
63. U. Fano and J.W. Cooper, Rev. Mod. Phys. 40, 441 (1968) [Russian translation].
64. J.P. Phillips, Spectra-Structure Correlation, Academic Press, New York (1964).
65. N.N. Kristofel' and G.S. Zavt, Opt. Spektrosk. 25, 705 (1968) [Opt. Spectrosc. (USSR) 25, 395 (1968)].
66. R. Boyn, J. Dziesiaty, and D. Wruck, Phys. Status Solidi B 42, K197 (1970).
67. J.M. Baranowski, J.M. Langer, and S. Stefanova, Proc. Eleventh Intern. Conf. on Physics of Semiconductors, Warsaw, 1972, Vol. 2, publ. by PWN, Warsaw (1972), p.1001.
68. S.A. Abagyan, G.A. Ivanov, Yu.N. Kuznetsov, Yu.A. Okunev, and Yu.E. Shanurin, Fiz. Tekh. Poluprovodn. 7, 1474 (1973) [Sov. Phys. Semicond. 7, 989 (1974)].
69. D. Bois and P. Pinard, Phys. Rev. B 9, 4171 (1974).
70. D.G. Andrianov, N.I. Suchkova, A.S. Savel'ev, E.P. Rashevskaya, and M.A. Filippov, Fiz. Tekh. Poluprovodn. 11, 730 (1977) [Sov. Phys. Semicond. 11, 426 (1977)].

71. N.I. Suchkova, D.G. Andrianov, É.M. Omel'yanovskiĭ, E.P. Rashevskaya, A.S. Savel'ev, V.I. Fistul', and M.A. Filippov, Fiz. Tekh. Poluprovodn. 11, 1742 (1977) [Sov. Phys. Semicond. 11, 1022 (1977)].
72. R. Bleekrode, J. Dieleman, and H.J. Vegter, Phys. Lett. 2, 355 (1962).
73. V.A. Presnov, V.K. Bazhenov, S.P. Fedotov, and B.S. Azikov, in: Chemical Bonds in Solids (ed. by N.N. Sirota), Vol. 1, General Problems in Electron Structure of Crystals, Consultants Bureau, New York (1972), p.81.
74. V.F. Masterov, B.E. Samorukov, K.F. Shtel'makh, and B.V. Chernovets, Izv. Vyssh. Uchebn. Zaved. Fiz. No. 9, 144 (1976).
75. A.V. Vasil'ev, G.K. Ippolitova, É.M. Omel'yanovskiĭ, and A.I. Ryskin, Fiz. Tekh. Poluprovodn. 10, 571 (1976) [Sov. Phys. Semicond. 10, 341 (1976)].
76. R.W. Haisty and G.R. Cronin, Proc. Seventh Intern. Conf. on Physics of Semiconductors, Paris, 1964, Vol. 1, Physics of Semiconductors, publ. by Dunod, Paris; Academic Press, New York (1964), p. 1161.
77. G.W. Ludwig and H.H. Woodbury, Bull. Am. Phys. Soc. 6, 118 (1961).
78. D.G. Andrianov, P.M. Grinshteĭn, G.K. Ippolitova, É.M. Omel'yanovskiĭ, N.I. Suchkova, and V.I. Fistul', Fiz. Tekh. Poluprovodn. 10, 1173 (1976) [Sov. Phys. Semicond. 10, 696 (1976)].
79. V.K. Sobolevskiĭ and K.F. Shtel'makh, in: Physics of III-V Compounds [in Russian], Leningrad Polytechnic Institute (1979), p. 66.
80. K. Suto and J. Nishizawa, J. Phys. Soc. Jpn. 26, 1556 (1969).
81. R.S. Title and T.S. Plaskett, Appl. Phys. Lett. 14, 76 (1969).
82. V.F. Masterov, B.E. Samorukov, V.K. Sobolevskiĭ, and K.F. Shtel'makh, Fiz. Tekh. Poluprovodn. 12, 529 (1978) [Sov. Phys. Semicond. 12, 305 (1978)].
83. S.A. Abagyan, G.A. Ivanov, Yu.N. Kuznetsov, and Yu.A. Okunev, Fiz. Tekh. Poluprovodn. 8, 1691 (1974) [Sov. Phys. Semicond. 8, 1096 (1975)].
84. S.A. Abagyan, G.A. Ivanov, G.A. Koroleva, Yu.N. Kuznetsov, and Yu.A. Okunev, Fiz. Tekh. Poluprovodn. 9, 369 (1975) [Sov. Phys. Semicond. 9, 243 (1975)].
85. K. Suto and J. Nishizawa, J. Appl. Phys. 43, 2247 (1972).
86. S.A. Abagyan, G.A. Ivanov, and G.A. Koroleva, Fiz. Tekh. Poluprovodn. 10, 1773 (1976) [Sov. Phys. Semicond. 10, 1056 (1976)].
87. G.K. Ippolitova, É.M. Omel'yanovskiĭ, N.M. Pavlov, A.Ya. Nashel'skiĭ, and S.V. Yakobson, Fiz. Tekh. Poluprovodn. 11, 1315 (1977) [Sov. Phys. Semicond. 11, 773 (1977)].
88. V.F. Masterov, B.E. Samorukov, and V.K. Sobolevskiĭ, Fiz. Tekh. Poluprovodn. 12, 201 (1978) [Sov. Phys. Semicond. 12, 117 (1978)].
89. T.L. Estle, Phys. Rev. 136, A1702 (1964).
90. J.M. Luttinger, Phys. Rev. 102, 1030 (1956).
91. É.M. Omel'yanovskiĭ, V.I. Fistul', L.A. Belagurov, V.S. Ivleva, V.V. Karataev, M.G. Mil'vidskiĭ, and A.N. Popkov, Fiz. Tekh. Poluprovodn. 9, 576 (1975) [Sov. Phys. Semicond. 9, 381 (1975)].
92. V.V. Kosarev and V.V. Popov, Abstracts of Papers presented at Fourth All-Union Conf. on Physicochemical Basis of Doping of Semiconductor Materials, Moscow 1979 [in Russian], A.A. Baĭkov Institute of Metallurgy, Academy of Sciences of the USSR, Moscow (1979), p. 14.
93. G.G.P. Van Gorkom and A.T. Vink, Solid State Commun. 11, 767 (1972).
94. K.H. Segsa and S. Spenke, Phys. Status Solidi A 27, 129 (1975).
95. V.F. Masterov and B.P. Popov, Fiz. Tekh. Poluprovodn. 12, 404 (1978) [Sov. Phys. Semicond. 12, 234 (1978)]; N.P. Il'in and V.F. Masterov, Fiz. Tekh. Poluprovodn. 12, 772 (1978) [Sov. Phys. Semicond. 12, 451 (1978)].
96. H. Watanabe, J. Phys. Chem. Solids 25, 1471 (1964).
97. R.S. Title, in: Physics and Chemistry of II-VI Compounds (ed. by M. Aven and J.S. Prener), North-Holland, Amsterdam (1967), p. 267.
98. B. Haydock, V. Heine, and M.J. Kelly, J. Phys. C 8, 2591 (1975).

99. W. Kohn and J.M. Luttinger, Phys. Rev. 98, 915 (1955).
100. W. Kohn, Solid State Phys. 5, 257 (1957).
101. S. Pantelides, Rev. Mod. Phys. 50, No. 4, 797 (1978).
102. S. Balasubramanian, J. Phys. C. 1, 1774 (1968).
103. M. Jaros, J. Phys. C 4, 1162 (1971).
104. P.K. Katana, Sh.D. Tiron, N.V. Dernovich, and A.G. Cheban, in: Physics of Impurity Centres [in Russian], Izd. AN Est. SSR, Tallinn (1970), p.81.
105. A. Morita, M. Azuma, and H. Nara, J. Phys. Soc. Jpn. 17, 1570 (1962).
106. H. Nara, J. Phys. Soc. Jpn. 20, 778 (1965).
107. T.I. Kucher and K.B. Tolpygo, Fiz. Tekh. Poluprovodn. 1, 77 (1967) [Sov. Phys. Semicond. 1, 59 (1967)].
108. T.H. Ning and C.T. Sah, Phys. Rev. B 4, 3468 (1971).
109. A. Baldereschi, Phys. Rev. B 1 4673 (1970).
110. H. Kaplan, J. Phys. Chem. Solids 24, 1593 (1963).
111. G.L. Bir, Fiz. Tverd. Tela (Leningrad) 13, 460 (1971) [Sov. Phys. Solid State 13, 371 (1971)].
112. F. Bassani, G. Iadonisi, B. Preziozi, Rev. Prog. Phys. 37, 1099 (1974).
113. R.A. Faulkner, Phys. Rev. 184, 713 (1969).
114. S.T. Pantelides and C.T. Sah, Phys. Rev. B 10, 621 (1974).
115. M. Altarelli, W.Y. Hsu, and R.A. Sabatini, J. Phys. C 10, L605 (1977).
116. L.V. Keldysh, Zh. Eksp. Teor. Fiz. 45, 364 (1963) [Sov. Phys. JETP 18, 253 (1964)].
117. E.O. Kane, J. Phys. Chem. Solids 1, 249 (1957).
118. A.A. Kolesnikov, Izv. Vyssh. Uchebn. Zaved. Fiz. No. 10, 38 (1972).
119. V.I. Sheka and D.I. Sheka, Zh. Eksp. Teor. Fiz. 51, 1445 (1966) [Sov. Phys. JETP 24, 975 (1967)].
120. V.I. Sheka and D.I. Sheka, Fiz. Tverd. Tela (Leningrad) 11, 1092 (1969) [Sov. Phys. Solid State 11, 891 (1969)].
121. K.B. Tolpygo, Fiz. Tverd. Tela (Leningrad) 11, 2846 (1969) [Sov. Phys. Solid State 11, 2304 (1970)].
122. V.N. Ryabokon' and K.K. Svidzinskiĭ, Fiz. Tverd. Tela (Leningrad) 11, 585 (1969) [Sov. Phys. Solid State 11, 473 (1969)].
123. V.N. Ryabokon' and K.K. Svidzinskiĭ, Fiz. Tekh. Poluprovodn. 5, 1865 (1971) [Sov. Phys. Semicond. 5, 1361 (1972)].
124. J.C. Phillips and L. Kleinman, Phys. Rev. 116, 287, 880 (1959).
125. A. Glodeanu, Phys. Status Solidi 35, 481 (1969).
126. A. Glodeanu, Rev. Roum. Phys. 10, 371, 433, 741 (1965).
127. V.N. Fleurov and K.A. Kikoin, J. Phys. C 9, 1673 (1976).
128. S.T. Pantelides and C.T. Sah, Phys. Rev. B 10, 638 (1974).
129. I.V. Abarenkov and V. Heine, Philos. Mag. 12, 529 (1965).
130. P.W. Anderson, Phys. Rev. 124, 41 (1961).
131. M. Jaros, Phys. Status Solidi 36, 181 (1969).
132. J. Bernholc and S.T. Pantelides, Phys. Rev. B 15, 4935 (1977).
133. F.D.M. Haldane and P.W. Anderson, Phys. Rev. B 13, 2553 (1976).
134. K.A. Kikoin and V.N. Fleurov, J. Phys. C 10, 4295 (1977).
135. G.F. Koster and J.C. Slater, Phys. Rev. 95, 1167 (1954); 96 1208 (1954).
136. I.M. Lifshits, Zh. Eksp. Teor. Fiz. 17, 1017 (1947).
137. J. Callaway and A.J. Hughes, Phys. Rev. 156, 860 (1967).
138. J. Callaway and A.J. Hughes, Phys. Rev. 164, 1043 (1967).
139. R.A. Faulkner, Phys. Rev. 175, 991 (1968).
140. M. Lannoo and P. Lenglart, J. Phys. Chem. Solids 30, 2409 (1969).
141. J. Bernholc and S.T. Pantelides, Phys. Rev. B 18, 1780 (1978).
142. R. Haydock, V. Heine, and M.J. Kelly, J. Phys. C 5, 2845 (1972).
143. N.P. Il'in and V.F. Masterov, Fiz. Tekh. Poluprovodn. 11, 1470 (1977) [Sov. Phys. Semicond. 11, 864 (1977)].

144. V.F. Masterov and B.E. Samorukov, Fiz. Tekh. Poluprovodn. 12, 625 (1978) [Sov. Phys. Semicond. 12, 363 (1978)].
145. J. Friedel, Adv. Phys. 3, 446 (1954).
146. C.A. Coulson and M.J. Kearsley, Proc. R. Soc. London Ser. A 241, 433 (1957).
147. T. Yamaguchi, J. Phys. Soc. Jpn. 17, 1359 (1962).
148. G.D. Watkins, Proc. Seventh Intern. Conf. on Physics of Semiconductors, Paris, 1964, Vol. 3, Radiation Damage in Semiconductors, publ. by Dunod, Paris; Academic Press, New York (1965), p. 97.
149. J. Friedel, M. Lannoo, and G. Leman, Phys. Rev. 164, 1056 (1967).
150. R. Messmer and G.D. Watkins, Phys. Rev. B 7, 2568 (1973).
151. F.P. Larkins, J. Phys. C 4, 3065 (1971).
152. M. Astier, N. Pottier, and J.C. Bourgoin, Phys. Rev. B 19, 5265 (1979).
153. C.M. Müller and U. Scherz, Phys. Rev. B 21, 5717 (1980).
154. C.M. Müller, U. Scherz, and C.M. Liegener, Proc. Fifteenth Intern. Conf. on Physics of Semiconductors, Kyoto, 1980, in: J. Phys. Soc. Jpn. 49, Suppl. A, 275 (1980).
155. B. G. Cartling, J. Phys. C 8, 3171, 3183 (1975).
156. L.A. Hemstreet, Phys. Rev. B 15, 834 (1977).
157. A. Fazzio and J.R. Leite, Phys. Rev. B 21, 4710 (1980).
158. J.C. Phillips, Phys. Rev. C 1, 1540 (1970).
159. G. Lucovsky, Solid State Commun. 3, 299 (1965).
160. S. Braun and H.G. Grimmeiss, J. Appl. Phys. 45, 2658 (1974).
161. S. Braun and H.G. Grimmeiss, Solid State Commun. 12, 657 (1973).
162. H.B. Bebb, Phys. Rev. 185, 1116 (1969).
163. J.W. Allen, J. Phys. C 2, 1077 (1969).
164. M. Jaros, J. Phys. C 4, 2979 (1971); 8, L264 (1975).
165. J.M. Baranowski and J.M. Langer, Phys. Status Solidi B 48, 863 (1971).
166. A.S. Baltenkov and A.A. Grinberg, Fiz. Tekh. Poluprovodn. 10, 1159 (1976) [Sov. Phys. Semicond. 10, 688 (1976)].
167. V.N. Flerov (Fleurov) and K.A. Kikoin, J. Phys. C 15, 3523 (1982).
168. B.K. Ridley, J. Phys. C 13, 2015 (1980).
169. A.B. Roĭtsin, Fiz. Tekh. Poluprovodn. 8, 3 (1974) [Sov. Phys. Semicond. 8, 1 (1974)].
170. S.T. Pantelides, J. Bernholc, and N.O. Lipari, Proc. Fifteenth Intern. Conf. on Physics of Semiconductors, Kyoto, 1980, in: J. Phys. Soc. Jpn. 49, Suppl. A, 235 (1980).
171. H.H. Woodbury and W.W. Tyler, Phys. Rev. 100, 659 (1955).
172. R. Newman and W.W. Tyler, Phys. Rev. 96, 882 (1954).
173. W.W. Tyler, R. Newman, and H.H. Woodbury, Phys. Rev. 97, 669 (1955).
174. W.W. Tyler, R. Newman, and H.H. Woodbury, Phys. Rev. 98, 461 (1955).
175. R. Newman and W.W. Tyler, Solid State Phys. 8, 49 (1959).
176. W.W. Tyler, J. Phys. Chem. Solids 8, 59 (1959).
177. A.G. Milnes, Deep Impurities in Semiconductors, Wiley, New York (1973).
178. M.L. Schultz, W.E. Harty, and C.D. Rowby, Report, RCA Laboratories, Princeton, N.J. (1962).
179. M.I. Barnik, B.I. Beglov, D.A. Romanychev, and Yu.S. Kharionovskiĭ, Fiz. Tekh. Poluprovodn. 5, 106 (1971) [Sov. Phys. Semicond. 5, 87 (1971)].
180. V.V. Ostroborodova and S.V. Ivanova, Fiz. Tverd. Tela (Leningrad) 6, 3481 (1964) [Sov. Phys. Solid State 6, 2787 (1965)].
181. M.M. Traum, M.J. Zucker, and S.H. Charap, Bull. Am. Phys. Soc. 14, 352 (1969).
182. K.D. Glinchuk, in: Current Topics in Physics of Semiconductors and Semiconductor Devices [in Russian], Vilnius (1969), p. 99.
183. E.H. Tyler, M. Jaros, and C.M. Penchina, Appl. Phys. Lett. 31, 208 (1977).

184. R.A. Malinauskas, L.Ya. Pervova, and V.I. Fistul', Fiz. Tekh. Poluprovodn. 13, 2270 (1979) [Sov. Phys. Semicond. 13, 1330 (1979)].
185. A.A. Kopylov and A.N. Pikhtin, Fiz. Tekh. Poluprovodn. 10, 15 (1976) [Sov. Phys. Semicond. 10, 7 (1976)].
186. N.T. Bendik, V.S. Garnyk, and L.S. Milevskiĭ, Fiz. Tverd. Tela (Leningrad) 12, 190 (1970) [Sov. Phys. Solid State 12, 150 (1970)].
187. A.A. Zolotukhin and L.S. Milevskiĭ, Fiz. Tverd. Tela (Leningrad) 13, 1906 (1971) [Sov. Phys. Solid State 13, 1598 (1972)].
188. R.O. Carlson, Phys. Rev. 104, 937 (1956).
189. M.K. Bakhadyrkhanov, S. Zaĭnabidinov, T.S. Kamilov, and A.T. Teshabaev, Fiz. Tekh. Poluprovodn. 8, 2263 (1974); 9, 76 (1975) [Sov. Phys. Semicond. 8, 1477 (1975); 9, 48 (1975)].
190. M.M. Akhmedova, L.S. Berman, L.S. Kostina, and A.A. Lebedev, Fiz. Tekh. Poluprovodn. 9, 2351 (1975) [Sov. Phys. Semicond. 9, 1516 (1975)].
191. C.B. Collins and R.O. Carlson, Phys. Rev. 108, 1409 (1957).
192. W.H. Shepherd and J.A. Turner, J. Phys. Chem. Solids 23, 1697 (1962).
193. N.T. Bendik, V.S. Garnyk, and L.S. Milevskiĭ, Fiz. Tverd. Tela (Leningrad) 12, 1693 (1970) [Sov. Phys. Solid State 12, 1340 (1970)].
194. B.I. Boltaks, M.K. Bakhadyrkhanov, and G.S. Kulikov, Fiz. Tverd. Tela (Leningrad) 13, 2675 (1971) [Sov. Phys. Solid State 13, 2240 (1972)].
195. A.A. Zolotukhin, A.K. Kovalenko, and L.S. Milevskiĭ, Fiz. Tverd. Tela (Leningrad) 13, 3119 (1971) [Sov. Phys. Solid State 13, 2621 (1972)].
196. H. Irmler, Z. Naturforsch. Teil A 13, 557 (1958).
197. H. Holonyak Jr, Proc. IRE 50, 2421 (1962).
198. S.K. Ghandhi, K.E. Mortenson, and J.N. Park, Proc. IRE 53, 635 (1965).
199. C.M. Penchina, J.S. Moore, and N. Holonyak Jr, Phys. Rev. 143, 634 (1966).
200. J.S. Moore, M.C.P. Chang, and C.M. Penchina, J. Appl. Phys. 41, 5282 (1970).
201. M.K. Bakhadyrkhanov, B.I. Boltaks, T.D. Dzhafarov, and G.S. Kulikov, Fiz. Tverd. Tela (Leningrad) 11, 3660 (1969) [Sov. Phys. Solid State 11, 3076 (1970)].
202. M.K. Bakhadyrkhanov, B.I. Boltaks, and G.S. Kulikov, Fiz. Tverd. Tela (Leningrad) 12, 181 (1970) [Sov. Phys. Solid State 12, 144 (1970)].
203. L.D. Yau, W.W. Chan, and C.T. Sah, Phys. Status Solidi A 14, 655 (1972).
204. M.K. Bakhadyrkhanov, S.S. Nigmankhodzhaev, and A.T. Teshabaev, Fiz. Tekh. Poluprovodn. 10, 606 (1976) [Sov. Phys. Semicond. 10, 364 (1976)].
205. Y. Tokumaru, Jpn. J. Appl. Phys. 2, 542 (1963).
206. S.K. Ghandhi, K.E. Mortenson, and J.M. Park, IEEE Trans. Electron Devices ED-13, 515 (1966).
207. W.B. Chua and K. Rose, J. Appl. Phys. 41, 2644 (1970).
208. B.I. Boltaks, M.K. Bakhadyrkhanov, S.M. Gorodetskiĭ, and G.S. Kulikov, Compensated Silicon [in Russian], Nauka, Leningrad (1972).
209. M. Yoshida and K. Saito, Jpn. J. Appl. Phys. 6, 573 (1967).
210. T. Tanaka and Y. Inuishi, J. Phys. Soc. Jpn. 19, 167 (1964).
211. F.P. Kesamanly and D.N. Nasledov, Gallium Arsenide: Preparation and Properties [in Russian], Nauka, Moscow (1973).
212. F.A. Cunnell, J.T. Edmond, and W.R. Harding, Solid-State Electron. 1, 97 (1960).
213. R.W. Haisty, Appl. Phys. Lett. 7, 208 (1965).
214. G.A. Allen, Br. J. Appl. Phys. 1, 593 (1968).
215. W. Plesiewicz, Phys. Status Solidi A 16, 485 (1973).
216. A.N. Imenkov, L.M. Kogan, M.M. Kozlov, S.S. Meskin, D.N. Nasledov, and B.V. Tsarenkov, Fiz. Tverd. Tela (Leningrad) 7, 3115 (1965) [Sov. Phys. Solid State 7, 2519 (1966)].
217. A. Kul'sreshta and A.É. Yunovich, Fiz. Tverd. Tela (Leningrad) 7, 2549 (1965) [Sov. Phys. Solid State 7, 2058 (1966)].

218. V.I. Budyanskiĭ, G.A. Shepel'skiĭ, N.M. Krolevets, E.A. Salkov, and M.K. Sheĭnkman, Ukr. Fiz. Zh. 17, 1918 (1972).
219. K. Kitahara, K. Nakai, M. Ozeki, A. Shibatomi, and K. Dazai, Jpn. J. Appl. Phys. 15, 2275 (1976).
220. V.K. Bazhenov, E.P. Rashevskaya, N.N. Solov'ev, and M.G. Foĭgel', Fiz. Tekh. Poluprovodn. 7, 1601 (1973) [Sov. Phys. Semicond. 7, 1067 (1974)].
221. E.W. Williams, Br. J. Appl. Phys. 18, 253 (1967).
222. H. Strack, Trans. Met. Soc. AIME 239, 381 (1967).
223. D.S. Domanevskiĭ, and V.D. Tkachev, Fiz. Tekh. Poluprovodn. 1, 377 (1967) [Sov. Phys. Semicond. 1, 310 (1967)].
224. É.M. Omel'yanovskiĭ, L.Ya. Pervova, E.P. Rashevskaya, N.N. Solov'ev, and V.I. Fistul', Fiz. Tekh. Poluprovodn. 4, 380 (1970) [Sov. Phys. Semicond. 4, 316 (1970)].
225. É.M. Omel'yanovskiĭ, N.M. Pavlov, N.N. Solov'ev, and V.I. Fistul', Fiz. Tekh. Poluprovodn. 4, 527 (1970) [Sov. Phys. Semicond. 4, 439 (1970)].
226. P.L. Hoyt and R.W. Haisty, J. Electrochem. Soc. 113, 296 (1966).
227. N.M. Kolchanova, Abstracts of Papers presented at Second All-Union Conf. on Deep Levels in Semiconductors, Tashkent, 1980 [in Russian], Part II, Tashkent State University (1980), p.25.
228. D.G. Andrianov, É.M. Omel'yanovskiĭ, E.P. Rashevskaya, and N.I. Suchkova, Fiz. Tekh. Poluprovodn. 10, 1071 (1976) [Sov. Phys. Semicond. 10, 637 (1976)].
229. L.N. Kurbatov, N.N. Mochalkin, A.D. Britov, É.M. Omel'yanovskiĭ, and N.N. Solov'ev, Fiz. Tekh. Poluprovodn. 3, 620 (1969) [Sov. Phys. Semicond. 3, 528 (1969)].
230. C.E. Jones Jr and A.R. Hilton, J. Electrochem. Soc. 113, 504 (1966).
231. D.C. Look, J. Phys. Chem. Solids 36, 1311 (1975).
232. L.A. Balagurov, É.M. Omel'yanovskiĭ, and L.Ya. Pervova, Fiz. Tekh. Poluprovodn. 8, 1616 (1974) [Sov. Phys. Semicond. 8, 1051 (1975)].
233. V.M. Gontar', G.A. Egizaryan, V.S. Rubin, V.I. Murygin, and V.I. Stafeev, Fiz. Tekh. Poluprovodn. 5, 1061 (1971) [Sov. Phys. Semicond. 5, 939 (1971)].
234. G.P. Peka and Yu.I. Karkhanin, Fiz. Tekh. Poluprovodn. 6, 305 (1972) [Sov. Phys. Semicond. 6, 261 (1972)].
235. A.G. Gorelenok, B.V. Tsarenkov, and N.G. Chiabrishvili, Fiz. Tekh. Poluprovodn. 4, 2064 (1970) [Sov. Phys. Semicond. 4, 1775 (1971)].
236. A.A. Gutkin, A.A. Lebedev, G.N. Talalakin, and T.A. Shaposhnikova, Fiz. Tekh. Poluprovodn. 6, 1067 (1972) [Sov. Phys. Semicond. 6, 928 (1972)].
237. N.M. Kalchanova, D.N. Nasledov, M.A. Mirdzhalilova, and V.Yu. Ibragimov, Fiz. Tekh. Poluprovodn. 4, 358 (1970) [Sov. Phys. Semicond. 4, 294 (1970)].
238. A.L. Lin and R.H. Bube, J. Appl. Phys. 47, 1859 (1976).
239. T. Nishino, T. Yanagida, and Y. Hamakawa, Jpn. J. Appl. Phys. 11, 1221 (1972).
240. W.J. Turner and G.D. Pettit, Bull. Am. Phys. Soc. 9, 269 (1964).
241. É.M. Omel'yanovskiĭ, L.Ya. Pervova, E.P. Rashevskaya, and V.I. Fistul', Fiz. Tekh. Poluprovodn. 5, 554 (1971) [Sov. Phys. Semicond. 5, 484 (1971)].
242. É.M. Omel'yanovskiĭ, A.N. Pantyukhov, L.Ya. Pervova, V.I. Fistul', and Yu.A. Vasil'ev, Fiz. Tekh. Poluprovodn. 9, 1930 (1975) [Sov. Phys. Semicond. 9, 1267 (1975)].
243. M.A. Messerer, É.M. Omel'yanovskiĭ, L.Ya. Pervova, Yu.Ya. Tkach, and V.I. Fistul', Fiz. Tekh. Poluprovodn. 10, 851 (1976) [Sov. Phys. Semicond. 10, 504 (1976)].
244. A.T. Gorelenok, B.V. Tsarenkov, and N.G. Chiabrishvili, Fiz. Tekh. Poluprovodn. 5, 115 (1971) [Sov. Phys. Semicond. 5, 95 (1971)].
245. N.V. Vorob'eva, Yu.V. Vorob'ev, and I.A. Kolomiets, Fiz. Tekh. Poluprovodn. 8, 606 (1974) [Sov. Phys. Semicond. 8, 388 (1974)].

246. H.J. Stocker and M. Schmidt, Proc. Thirteenth Intern. Conf. on Physics of Semiconductors, Rome, 1976, publ. by North-Holland, Amsterdam (1976), p. 611.
247. W.J. Turner, G.D. Pettit, and N.G. Ainslie, J. Appl. Phys. 34, 3274 (1963).
248. L.R. Weisberg, F.D. Rosi, and P.G. Herkart, in: Properties of Elemental and Compound Semiconductors (Proc. AIME, Boston, 1959, ed. by H.C. Gatos), Interscience, New York (1960), p. 25.
249. V.I. Ivanenko, É.A. Poltoratskiĭ, V.S. Rubin, and A.V. Strieva, Fiz. Tekh. Poluprovodn. 5, 1235 (1971) [Sov. Phys. Semicond. 5, 1086 (1971)].
250. Yu.A. Matveenko, V.I. Murygin, and V.S. Rubin, Fiz. Tekh. Poluprovodn. 2, 1844 (1968) [Sov. Phys. Semicond. 2, 1536 (1969)].
251. W.J. Brown Jr and J.S. Blakemore, J. Appl. Phys. 43, 2242 (1972).
252. V.I. Murygin and V.S. Rubin, Fiz. Tekh. Poluprovodn. 3, 959 (1969) [Sov. Phys. Semicond. 3, 810 (1970)].
253. N.M. Kolchanova, D.N. Nasledov, and G.N. Talalakin, Fiz. Tekh. Poluprovodn. 4, 134 (1970) [Sov. Phys. Semicond. 4, 106 (1970)].
254. V.I. Fistul' and A.M. Agaev, Fiz. Tverd. Tela (Leningrad) 7, 3681 (1965) [Sov. Phys. Solid State 7, 2988 (1966)].
255. J.M. Baranowski, M. Grynberg, and E.M. Magerramov, Phys. Status Solidi B, 50, 433 (1972).
256. M.F. Bekmuratov and V.I. Murygin, Fiz. Tekh. Poluprovodn. 7, 83 (1973) [Sov. Phys. Semicond. 7, 55 (1973)].
257. D.G. Andrianov, A.S. Savel'ev, N.I. Suchkova, E.P. Rashevskaya, and M.A. Filippov, Fiz. Tekh. Poluprovodn. 11, 1460 (1977) [Sov. Phys. Semicond. 11, 858 (1977)].
258. F. Willmann, M. Blatte, H.J. Queisser, and J. Treusch, Solid State Commun. 9, 2281 (1971).
259. A.V. Vasil'ev, G.K. Ippolitova, É.M. Omel'yanovskiĭ, and A.I. Ryskin, Fiz. Tekh. Poluprovodn. 10, 571 (1976) [Sov. Phys. Semicond. 10, 341 (1976)].
260. S.A. Abagyan, G.A. Ivanov, and Yu.N. Kuznetsov, Fiz. Tekh. Poluprovodn. 10, 2160 (1976) [Sov. Phys. Semicond. 10, 1283 (1976)].
261. D.H. Loescher, J.W. Allen, and G.L. Pearson, Proc. Eighth Intern. Conf. on Physics of Semiconductors, Kyoto, 1966, in: J. Phys. Soc. Jpn. 21, Suppl., 239 (1966).
262. Y. Okuno, K. Suto, and J. Nishizawa, J. Appl. Phys. 44, 832 (1973).
263. F.S. Shishiyanu and V.G. Georgiu, Fiz. Tekh. Poluprovodn. 10, 2188 (1976) [Sov. Phys. Semicond. 10, 1301 (1976)].
264. G.G. Kovalevskaya, V.I. Alyushina, and S.V. Slobodchikov, Fiz. Tekh. Poluprovodn. 9, 2125 (1975) [Sov. Phys. Semicond. 9, 1385 (1975)].
265. N.S. Grushko, É.V. Russu, and S.V. Slobodchikov, Fiz. Tekh. Poluprovodn. 9, 343 (1975) [Sov. Phys. Semicond. 9, 224 (1975)].
266. N.S. Grushko and A.A. Gutkin, Fiz. Tekh. Poluprovodn. 8, 1816 (1974) [Sov. Phys. Semicond. 8, 1179 (1975)].
267. S.V. Bulyarskiĭ, N.S. Grushko, G.S. Korotchenkov, and I.P. Molodyan, Deposited Paper No. 6668-73 [in Russian], VINITI, Moscow (1973).
268. S.V. Bulyarskiĭ, N.S. Grushko, A.A. Gutkin, and D.N. Nasledov, Fiz. Tekh. Poluprovodn. 9, 287 (1975) [Sov. Phys. Semicond. 9, 187 (1975)].
269. S.P. Starosel'tseva, V.S. Kulov, and S.G. Metreveli, Fiz. Tekh. Poluprovodn. 5, 1842 (1971) [Sov. Phys. Semicond. 5, 1603 (1972)].
270. V.I. Petrovskiĭ, N.N. Solov'ev, É.M. Omel'yanovskiĭ, and V.S. Ivleva, Fiz. Tekh. Poluprovodn. 12, 1904 (1978) [Sov. Phys. Semicond. 12, 1132 (1978)].
271. L.A. Balagurov, É.M. Omel'yanovskiĭ, and V.I. Fistul', Fiz. Tekh. Poluprovodn. 12, 944 (1978) [Sov. Phys. Semicond. 12, 557 (1978)].
272. Ya.G. Dorfman, Diamagnetism and the Chemical Bond, American Elsevier, New York (1965).

273. N.N. Sirota, in: Chemical Bonds in Semiconductors and Solids [in Russian], Nauka i Tekhnika, Minsk (1965), p. 35.
274. D.G. Andrianov, L.A. Zhukova, A.S. Savel'ev, and V.I. Fistul', Kristallografiya 19, 802 (1974) [Sov. Phys. Crystallogr. 19, 497 (1975)].
275. W. Zawadzki, Phys. Status Solidi 3, 1421 (1963).
276. J. Kolodziejczak and L. Sosnowski, Acta Phys. Pol. 21, 399 (1962).
277. A.S. Savel'ev and D.G. Andrianov, Fiz. Tekh. Poluprovodn. 6, 404 (1972) [Sov. Phys. Semicond. 6, 347 (1972)].
278. R. Kowalczyk, K. Kolodziejczak, and W. Zawadzki, The Generalized Fermi–Dirac Integrals, Polish Academy of Sciences, Warsaw (1965).
279. E. Mooser, Phys. Rev. 100, 1589 (1955).
280. E. Sonder and H.C. Schweinler, Phys. Rev. 117, 1216 (1960).
281. B.I. Shklovskii and A.L. Efros, Electronic Properties of Doped Semiconductors, Springer Verlag, Berlin (1984).
282. G.V. Il'menkov, D.N. Nasledov, Yu.S. Smetannikova, and V.R. Felitsiant, Fiz. Tverd. Tela (Leningrad) 11, 3282 (1969) [Sov. Phys. Solid State 11, 2659 (1970)].
283. G.A. Slack, S. Roberts, and J. Vallin, Phys. Rev. 187, 511 (1969).
284. D.G. Andrianov, A.S. Savel'ev, and V.I. Fistul', Fiz. Tekh. Poluprovodn. 9, 136 (1975) [Sov. Phys. Semicond. 9, 563 (1975)].
285. D. Geist, Z. Phys. 158, 123 (1960).
286. H. Wagini and M. Wilhelm, Z. Naturforsch. Teil A 21, 329 (1966).
287. D.G. Andrianov, Yu. B. Muravlev, A.S. Savel'ev, N.N. Solov'ev, and V.I. Fistul', Fiz. Tekh. Poluprovodn. 7, 1622 (1973) [Sov. Phys. Semicond. 7, 1083 (1974)].
288. W. Low and M. Weger, Phys. Rev. 118, 1119 (1960).
289. D.G. Andrianov, A.S. Savel'ev, and V.I. Fistul', Fiz. Tekh. Poluprovodn. 9, 136 (1975) [Sov. Phys. Semicond. 9, 89 (1975)].
290. J.P. Mahoney, C.C. Lin, and W.H. Brumage, J. Chem. Phys. 50, 2263 (1969).
291. D.G. Andrianov and A.S. Savel'ev, Fiz. Tekh. Poluprovodn. 14, 539 (1980) [Sov. Phys. Semicond. 14, 317 (1980)].
292. J.T. Vallin, G.A. Slack, S. Roberts, and A.E. Hughes, Phys. Rev. B 2, 4313 (1970).
293. D.G. Andrianov, G.V. Lazareva, A.S. Savel'ev, V.I. Selyanina, and V.I. Fistul', Fiz. Tekh. Poluprovodn. 9, 1555 (1975) [Sov. Phys. Semicond. 9, 1026 (1975)].
294. Y. Toyozawa, J. Phys. Soc. Jpn. 17, 986 (1962).
295. K. Yosida, Phys. Rev. 107, 396 (1957).
296. Yu.V. Shmartsev, E.F. Shender, and T.A. Polyanskaya, Fiz. Tekh. Poluprovodn. 4, 2311 (1970) [Sov. Phys. Semicond. 4, 1990 (1971)].
297. G. Garyagdyev, O.V. Emel'yanenko, N.V. Zotova, T.S. Lagunova, and D.N. Nasledov, Fiz. Tekh. Poloprovodn. 7, 700 (1973) [Sov. Phys. Semicond. 7, 487 (1973)].
298. Sh.M. Gasanli, O.V. Emel'yanenko, T.S. Lagunova, and D.N. Nasledov, Fiz. Tekh. Poluprovodn. 6, 2010 (1972) [Sov. Phys. Semicond. 6, 1714 (1973)].
299. É.I. Zavaritskaya, I.D. Voronova, and N.V. Rozhdestvenskaya, Fiz. Tekh. Poluprovodn. 6, 1945 (1972) [Sov. Phys. Semicond. 6, 1668 (1973)].
300. D.G. Andrianov, E.P. Rashevskaya, and V.I. Fistul', Fiz. Tekh. Poluprovodn. 1, 1435 (1967) [Sov. Phys. Semicond. 1, 1195 (1968)].
301. G. Krill and M.F. Lapierre, Phys. Lett. A 35, 301 (1971).
302. D.G. Andrianov, G.V. Lazareva, A.S. Savel'ev, and V.I. Fistul', Fiz. Tekh. Poluprovodn. 9, 210 (1975) [Sov. Phys. Semicond. 9, 141 (1975)].
303. B. Pödör, Phys. Status Solidi 31, K55 (1969).
304. R.P. Khosla and J.R. Fischer, Phys. Rev. B 2, 4084 (1970).
304a. B.L. Al'tshuler, A.G. Aronov, A.I. Larkin, and D.E. Khmel'nitskiĭ, Zh. Eksp. Teor. Fiz. 81, 768 (1981) [Sov. Phys. JETP 54, 411 (1981)].
305. D.G. Andrianov, G.V. Lazareva, A.S. Savel'ev, and V.I. Fistul', Fiz. Tekh. Poluprovodn. 10, 568 (1976) [Sov. Phys. Semicond. 10, 339 (1976)].
306. W. Plesiewicz, Phys. Status Solidi A 16, 485 (1973).

307. L. Gouskov, S. Bilac, J. Pimentel, and A. Gouskov, Solid-State Electron. 20, 653 (1977).
308. O. Mizuno, S. Kikuchi, and Y. Seki, Jpn. J. Appl. Phys. 10, 208 (1971).
309. M. Otsubo and H. Miki, J. Electrochem. Soc. 124, 441 (1977).
310. G. Picoli, B. Deveaud, and D. Galland, J. Phys. (Paris) 42, 133 (1981).
311. A.M. White, Solid State Commun. 32, 205 (1979).

ADDITIONAL REFERENCES FOR THE ENGLISH EDITION

1. P.J. Williams, L. Eaves, P.E. Simmonds, M.O. Henry, E.C. Lightowlers, and Ch. Uihlein, J. Phys. C 15, 1337 (1982).
2. L. Eaves, A.W. Smith, M.S. Skolnick, and B. Cockayne, J. Appl. Phys. 53, 4955 (1982).
3. Semi-Insulating III-V Materials (Proc. Intern. Conf., Evian, 1982, ed. by B. Tuck and S. Makram-Ebeid), Shiva Publishing, Nantwich, England (1982).
4. Semi-Insulating III-V Materials (Proc. Intern. Conf., Warm Springs, Oregon, 1984, ed. by D.C. Look and J.S. Blakemore), Shiva Publishing, Nantwich, England (1984).
5. U. Lindefelt and A. Zunger, Phys. Rev. B 26, 846 (1982).
6. A. Zunger and U. Lindefelt, Phys. Rev. B 27, 1191 (1983).
7. V.F. Masterov, Fiz. Tekh. Poluprovodn. 18, 3 (1984) [Sov. Phys. Semicond. 18, 1 (1984)].
8. M.S. Skolnick, E.J. Foulkes, and B. Tuck, J. Appl. Phys. 55, 2951 (1984).